Choosing & Raising
CHICKENS

Choosing & Raising
CHICKENS

The complete
guide to breeds
and welfare

JEREMY HOBSON & CELIA LEWIS

D&C
David and Charles

"For my father Brian – always an enthusiastic supporter
of all my interests and in particular, chicken keeping."

A DAVID & CHARLES BOOK
Copyright © David & Charles Limited 2009

David & Charles is an F+W Media, Inc. company
4700 East Galbraith Road
Cincinnati, OH 45236

First published in the UK in 2009

Text copyright © Jeremy Hobson and Celia Lewis 2009
Photography and artwork © David & Charles 2009, except those listed on page 160

Some material from this book has been previously published in Keeping Chickens,
published by David & Charles Ltd, 2007.

A catalogue record for this book is available from the British Library.

ISBN-13: 978-0-7153-3310-5 paperback
ISBN-10: 0-7153-3310-0 paperback

Printed in China by RR Donnelley
for David & Charles
Brunel House   Newton Abbot   Devon

Commissioning Editor: Neil Baber
Editorial Manager: Emily Pitcher
Editor: Verity Muir
Project Editor: Caroline Taggart
Art Editor: Sarah Clark
Design Assistant: Victoria Marks
Production Controller: Kelly Smith

Visit our website at www.davidandcharles.co.uk

David & Charles books are available from all good bookshops; alternatively you can contact
our Orderline on 0870 9908222 or write to us at FREEPOST EX2 110, D&C Direct,
Newton Abbot, TQ12 4ZZ (no stamp required UK only); US customers call 800-289-0963
and Canadian customers call 800-840-5220.

CONTENTS

Left: Chickens can successfully be kept in an average-sized garden, just so long as precautions are taken to keep them safe from the unwanted attentions of other family pets.

Below: A pair of Pekins find a spot of shade beneath a raised chicken house.

INTRODUCTION

A plot of land without chickens can be like an empty house, bereft of the hustle and bustle of its owners. Poultry can invigorate the lowliest environment or enhance the most beautiful. Picture a sun-lit garden full of blooming flowers. Now imagine a handsome cockerel strolling proudly across the grass with his harem of hens. The scene is transformed.

Chickens quite literally add 'life' to your life, providing a source of joy and relaxation after a stressful day. They are relatively inexpensive to feed and keep, will not take up too much of your time and come in such a range of shapes and sizes that they will fit in virtually anywhere. Best of all, your hens will supply you with the ingredients for many varied meals – what more could one ask for?

There is, however, no escaping the fact that people and poultry may, in certain circumstances, coexist uneasily, although with care and forethought most of the common problems can be avoided or overcome. Chickens and bantams can live alongside neighbours, herbaceous borders and a busy lifestyle, but it is important that the right breeds are chosen to match the situation. The correct numerical balance must also be struck, and the henhouse should be positioned so as not to cause offence. A guiding hand can often help with issues such as these but, in the absence of an experienced neighbour, an easily read book can be just as useful.

In the hope of answering poultry-related questions before they arise, *Choosing & Raising Chickens* explains both the theory and the practical aspects of chicken keeping. It also includes quirky 'I never knew that' pieces of information. Who would have thought, for instance, that the Scots Dumpy was bred with short legs to stop it wandering too far away from the croft, or that only white eggs were seen in the UK until the arrival of Asiatic breeds in the mid-19th century? And did you know that alektorophobia is the fear of chickens?

No matter how experienced a person may be in a certain area, he or she cannot know everything there is to know about it and will still have things to learn on a daily basis. Nowhere is this more true than when keeping chickens. This book is written by two experienced poultry breeders and authors, who hope both to introduce the delights of chicken keeping to the newcomer and to share their fascination in the hobby with those who are already seasoned campaigners.

Opposite: A pair of Barbu d'Uccle bantams make themselves at home on an unusual perch.

HOW TO USE THIS BOOK

This book is broadly split into three main areas of interest: what to consider before buying chickens; how to select the perfect breed for your needs; and the practicalities of chicken keeping. An indispensable glossary (pages 154–155) outlines all the terms you are most likely to encounter.

THE FACT FILE BOX

The essential information for each breed is presented in a useful display box. The headings featured are explained here.

BREED TYPE

Soft-feather: breeds with a profusion of feathering which can often disguise the shape and size of the body.

Hard-feather: breeds with close, tight feathering which follows the shape of the body.

Light: traditionally breeds with a good reputation for egg laying. Light breeds tend towards flightiness and it may be necessary to build taller sides to the run or to top it with nylon netting. They generally never become quite as tame as other breeds making them less suitable if you want a pet bird.

Heavy: historically breeds developed as table birds (produced for their meat). Generally large and heavy in stature they can be kept with the minimum amount of fencing, and their calm, placid nature makes them ideal pets and companions and great for beginners.

Rare: Rare breeds may not have their own specialist breed club and in the UK these are catered for by the Rare Poultry Society.

True bantam: naturally diminutive fowls with no large counterpart. They are mainly kept for ornamental purposes and for exhibitions, or as pets.

BANTAM

Most (but not all) fowl can be bred as miniatures generally one quarter of the size of the large breed, making them an obvious choice where space is limited. Some birds are true bantams (see above).

PURPOSE

Layer: a bird kept for egg production.
Table: developed for their meat.
Dual: breeds like the Rhode Island Red that are both meat birds and productive egg-layers.
Exhibition/Ornamental: some breeds are only suitable for ornamental purposes or for exhibition.

EGG LAYING

Some chickens produce more eggs than others and if you are looking for a year-round supply of eggs you will need to know how good a layer each breed is. Here is your tick tally:

√	= 0–100 eggs
√√	= 100–150 eggs
√√√	= 150–200 eggs
√√√√	= 200–250 eggs
√√√√√	= 250–300+ eggs

EASE OF KEEPING

Certain physical characteristics of chicken breeds make them more difficult to care for than others, for example the long tail feathers of the Yokohama will require specific housing needs. The more ticks a breed has the easier it is to keep.

SOCIABILITY

Some breeds are more sociable with each other than others. For example, the Aseel was bred as a fighting bird and both the male and female can be very aggressive. The more ticks a breed has the more sociable it is.

The Fact File information for each breed has been compiled into a very useful table which can be found on pages 152–153 of this book. This will help you to compare the relative merits of each of the featured breeds.

A   The breed name

B   Interesting information about the history, appearance and purpose of the breed

C   Identifying breed photographs (M=male/F=female)

D   Fact File box – the essential information at a glance (see opposite for more explanation)

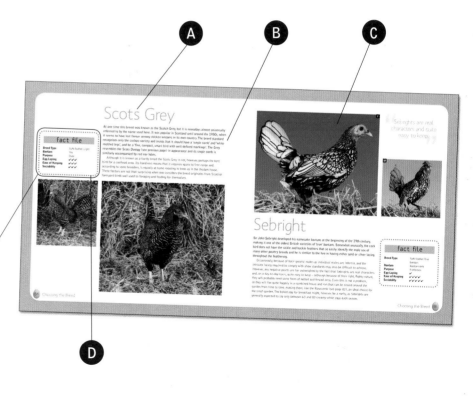

Scots Grey

At one time this breed was known as the Scotch Grey, but it is nowadays almost universally referred to by the name used here. It was popular in Scotland until around the 1930s, when it seems to have lost favour among chicken keepers in its own country. The breed standard recognizes only the cuckoo variety and insists that it should have a 'single comb' and 'white mottled legs', and be a 'fine, compact, smart bird with well-defined markings'. The Grey resembles the Scots Dorking (see previous page) in appearance and its single comb is similarly accompanied by red ear lobes.

Although it is known as a hardy breed the Scots Grey is not, however, perhaps the best bird for a confined area. Its hardiness means that it requires space to free-range and, according to some breeders, is equally at home roosting in trees as in the chicken house. These factors are not that surprising when one considers the breed originates from Scottish farmyard birds well used to foraging and feeding for themselves.

Sebrights are real characters and quite easy to keep.

Sebright

Sir John Sebright developed his namesake bantam at the beginning of the 19th century, making it one of the oldest British varieties of 'true' bantam. Somewhat unusually, the cock bird does not have the sickle and hackle feathers that so easily identify the male sex of many other poultry breeds and he is similar to the hen in having either gold or silver lacing throughout the feathering.

Occasionally, because of their genetic make-up individual males are infertile, and the intricate lacing required to comply with show standards may also be difficult to achieve. However, any negative points are far outweighed by the fact that Sebrights are real characters and, on a day-to-day basis, quite easy to keep – although because of their light, flighty nature they will probably need some form of netted and fenced area. Even this is not a problem, as they will fair quite happily in a combined house and run that can be moved around the garden from time to time, making them, like the Rosecomb (see page 82), an ideal choice for the small garden. The boiled egg for breakfast might, however, be a rarity, as Sebrights are generally expected to lay only between 60 and 80 creamy white eggs each season.

fact file

Breed Type	Soft-feather. Light
Bantam	Yes
Purpose	Dual
Egg Laying	✔✔✔
Ease of Keeping	✔✔✔
Sociability	✔✔✔

ANATOMY OF A CHICKEN

Poultry breeds vary widely in appearance, colouring and feather formation, but all share the basic elements of external anatomy (page 10).

COMBS, WATTLES AND EAR LOBES

One of the most distinctive features of a chicken is its comb, which, with its wattles (fleshy appendages under the beak), functions as its cooling system. Birds cannot sweat, so the chicken cools itself by circulating blood through its comb and wattles, radiating body heat.

There are several types of comb, the most common of which is the single comb. This is serrated and stands upright on the head, although in some breeds with very large single combs the back part tends to droop to one side. In contrast, the rose comb is low and broad, lacks points and tapers off into a 'spike' at the back. There are also V-shaped combs, walnut or cushion combs, pea combs and buttercup combs. Some breeds, such as the Poland, have only rudimentary or absent combs.

Surprisingly, ear-lobe colour governs egg colour: generally, if the hen has white ear lobes she will lay white-shelled eggs, whereas red ear lobes indicate brown eggs. One exception to this rule is the Araucana (see page 26), which has red ear lobes but lays blue- or green-shelled eggs.

FEATHERS

Chicken feathers come in many colours and patterns, and provide both insulation and waterproofing. The condition of a bird's feathers is maintained through preening, which takes up a large proportion of its day and involves dust-bathing and applying oil from a gland at the base of the tail. From a zoological point of view, feathers are composed of keratin, like hair and horn, and correspond to the scales of a fish. And like scales, they overlap one another.

Poultry breeds are categorized as hard- or soft-feathered: the former are those originally bred for cock fighting, which have smooth, close-fitting plumage. Not all birds have the same amount of feathering – some breeds have feathers right down their legs and may

also have feather beards and crests.

The hackle and saddle feathers of cock birds can be distinguished from those of hens as they tend to be pointed and shiny. The crest feathers of the cock are the same shape as its hackles, while in hens they are shorter and broader. Cocks also have sickles – long, curved feathers on either side of the tail.

Distinctive feather patterns can be used to identify the many varieties of chicken:
- **Barred** feathers have alternate transverse markings in two distinct colours.
- **Frizzled** feathers curl and curve outwards and forwards in a random manner.
- **Laced** feathers have a border of a contrasting colour all around the edge.
- **Mottled** feathers have a variable percentage of black feathers tipped with white.
- **Spangled** feathers have a distinct marking of a contrasting colour, usually in a V shape at their tip.
- **Pencilled** feathers include several types of lines or markings, although most commonly they have narrow concentric linear markings following the contours of the edges.

single comb

rose comb

pea comb

walnut comb

v-shaped comb

buttercup comb

| *barred* | *laced* | *pencilled* | *spangled* |

THE PECKING ORDER

Chickens are gregarious birds that live together as a flock, and the harmony of the group is maintained by the observation of a hierarchy, termed the pecking order, which governs where each chicken stands within the flock.

The pecking order starts with the 'top bird' and extends down to the youngest and weakest, which survive as best they can. The top bird is usually an old hen, although sometimes it is the most aggressive bird, and the rest of the flock will defer to her, often giving way at the food container.

When new birds are introduced to an established flock there are always problems because of the upset to the pecking order. A hen that spots a newcomer will utter a single warning croak, alerting the rest of the flock. If the birds were living in the wild the newcomers would be seen off; in a run they may be bullied unmercifully until they are eventually accepted. One of the points in favour of keeping a cockerel is that he will stop the older members of the flock bullying the junior hens. To help protect new birds from becoming henpecked in this way, introduce them at night and never one at a time.

Alternatively, if possible, introduce all the birds to one another by placing the newcomers in an adjoining coop and run, where they can be seen but are safe from attack behind wire netting. After a few days it will be safe to let them all intermingle – at first the new kids on the block will probably go back to their old roost each evening, but eventually they will join the majority in the main house.

Without an established flock to show them what to do, new birds can be remarkably stupid about knowing that they should go into their house at night and roost on the perches provided, often preferring instead to fly up into trees or to teeter on the henhouse roof. When you buy new stock, it is a good idea to make sure that all the birds are perching happily for the first few nights; if they are not, lift them on to their perches until they learn to do this themselves.

Contrary to popular belief, it is possible to run a large flock of birds containing two cockerels, but much depends on the breeds chosen (see the details of the individual breeds beginning on page 18). Some, usually the heavier, more docile ones, will share a harem of hens quite happily, whilst others, notably the lighter game breeds, will not tolerate any male competition once they are sexually mature.

The basic elements of external anatomy of a fowl.

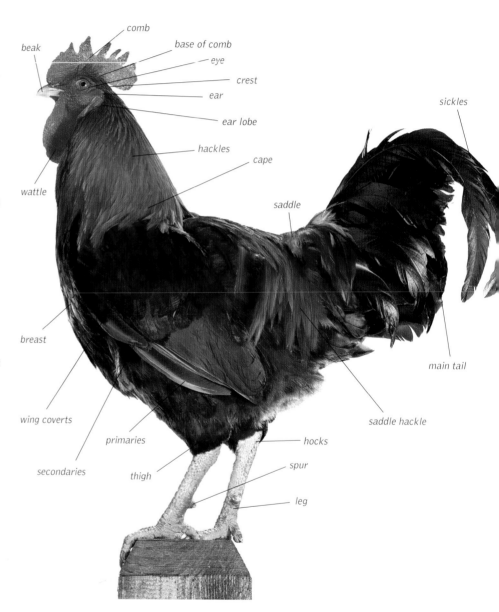

comb

base of comb

beak

eye

crest

ear

ear lobe

sickles

hackles

cape

wattle

saddle

breast

main tail

wing coverts

saddle hackle

primaries

hocks

secondaries

spur

thigh

leg

DECIDING TO KEEP CHICKENS

Before you decide on species and systems of management you need to outline your intentions – do you plan to keep chickens for profit or self sufficiency, or simply for pleasure? You should then evaluate how much time and space you will need. A small pen of bantams can be managed on a scaled operation, but larger, featherlegged breeds can be demanding and will require far greater attention.

SPACE

The theory behind keeping free-range poultry is wonderful but the practice is not always quite so simple. Although free-ranging poultry will eradicate insect pests and slugs in the garden, some of their habits are less welcome. Chickens and bantams love nothing more than a good dust-bath to help rid themselves of parasites, and consider a well-prepared seedbed ideal for the purpose. And to their mind, bark mulch that has been carefully placed around plants is scratching heaven. For these reasons alone, you may prefer to keep your chickens confined to a run. See page 112 for more about this.

CHILDREN AND PETS

While chickens make excellent pets for children (and keeping chickens is an excellent hobby for children), some breeds are known to be more aggressive or flighty than others. It may be sensible, especially if you have toddlers, to avoid keeping cock birds of these breeds. Instead, opt for bantams, whose smaller size is also more likely to appeal to youngsters.

If you already have other pets, consider how they are likely to react to the arrival of a flock of chickens in their territory. Most cats get on well with hens, but the behaviour of dogs varies according to temperament. Most will accept the flock as part of the family and show no interest in them, but others may attack and kill them – this is especially true with some types of bantam, which seem to have a more gamey smell. Rabbits and guinea pigs will live happily near hens but should not be put in with a flock that is fenced in a small run. Hens and ponies are an excellent mix – ponies will enjoy the company of the birds and the hens will scratch around, eating up all the dropped food and hay seed.

Free-ranging chickens can bring a garden to life, as long as you take precautions to stop them ruining your flower beds!

Keeping chickens is an excellent hobby for children, teaching them responsibility through having to care for other creatures.

DAILY ROUTINE

Although chickens and bantams are not time-consuming to look after, they do appreciate a routine and like to be let out of their house and fed at regular times. Keeping to a daily pattern is important, so before you decide to make any purchases it is vital that you consider how you will carry out the apparently simple tasks of shutting the chickens up at night and letting them out in the morning.

In the winter months in more northerly latitudes, many people leave for work in the dark and, therefore, before the birds come down off the roost. Will there be anyone at home to let your chickens out once daybreak comes? And, if so, can you rely on them not to forget? In the middle of summer, there may be problems at the other end of the day, as your chickens may still be scratching around when you are ready for bed. If the birds are kept in a secure pen this may not be a problem, but if they are free-range there is a strong possibility that they will be taken by a nocturnal predator.

Ideally, chickens should be fed little and often. In a busy household, however, this is not always practical, so feeding morning and evening is the best option. Don't be tempted to leave out too much food: not only is this wasteful, but pellets or mash can go sour in damp conditions and the practice will certainly encourage vermin. If food and water are given ad lib in the henhouse, the floor covering will get scratched into the feed and water will be spilt, increasing the likelihood of fungal disease.

Other parts of the daily routine include scraping the previous night's faeces from the droppings board, checking that no eggs have been broken in the nest boxes and changing damp litter around any indoor drinking vessels. In the winter, drinkers need to be kept free of ice. (For more information on cleaning and maintenance routines, see page 112.)

Although none of these chores takes too long, there is no point in keeping chickens if you cannot spend some time watching and enjoying their antics. In addition to providing relaxation at the end of a busy day, time spent leaning on the gate is essential in order to keep a check on your birds' well-being.

WHEN YOU GO AWAY

When you are starting out on the exciting adventure of keeping chickens, bear in mind that you will need other people's assistance when you go away. If you live among like-minded neighbours, it is not too onerous a task for one of them to pop round each morning and evening to feed and water your birds. However, if you live down a lonely track out in the country, finding someone to take on the job may not be so simple.

Equally, a small flock of birds is not likely to cause too much bother for anyone taking over their care, but if you have several different breeding pens, incubators and rearing cages it is probably too great a responsibility for a neighbour or friend. If you only ever have one annual holiday, you can arrange this at the least busy time in your poultry year, but if you go away regularly the size and demands of your flock should reflect this.

On this subject, also plan the pen and poultry shed layout with others in mind – a gate that falls off its hinges and traps your ankle every time you go through it might not bother you, but you can be sure that outside helpers will be a little less enthusiastic the next time you call on their assistance. You could also consider setting up an automatic watering system. There are several of these on the market that are inexpensive and easy to install, and they will save both you and your stand-ins some time. Commercially produced automatic pop-holes are also available, although if you are creative it is easy enough to design and build one yourself that is raised and lowered by the use of a small motor and timer. However, if you do install such a device you will still need someone to check that the system hasn't failed and that the birds haven't been locked in all day or, worse still, locked out at night.

CHICKENS & THE GARDEN

If you want to keep free-range chickens but at the same time would like to maintain a reasonably tidy garden, there are several tricks you can employ:

Choose bantams with feathered legs such as Pekins, as these don't, or can't, scratch up the grass on your lawn quite as much as other breeds.

•

Lay chicken wire flat on your herbaceous or annual beds in winter. The plants will grow up through the wire but the hens won't be able to scratch them up.

•

If dust-bathing in your beds becomes a problem, try tempting your birds away from them with an irresistible box of dry sand or peat. If you place it in a sunny position they may prefer it to the flower beds.

•

If your hens take to laying outside rather than in the cosy nest boxes you have provided, do not let them out of their run until mid-morning, by which time most should have finished laying. Some birds may, however, wait to visit their favourite spot, in which case you will have to spy on them, as they won't go near it if they know they are being watched.

•

Fertilize your plants with the manure your chickens produce – around 100g (4oz) per bird per day. It is high in nitrogen, which keeps foliage green and makes an excellent lawn feed when diluted in water. However, as it is high in ammonia it will burn plants if applied direct to beds. Instead, add it to your compost to allow it to break down first – it will also help to heat the heap and aid rotting.

Diluted in water, chicken droppings make an effective, nitrogen-rich lawn feed. Don't apply it direct to flower beds – it will burn the plants.

COST

Aside from the birds themselves, the initial costs of keeping chickens are those of the capital equipment – housing, wire netting, fence posts, feeders and drinkers. If you intend breeding more than the odd clutch of eggs, you might also consider the acquisition of a small incubator and artificial brooder.

Look after it well and there is no reason why such equipment should not remain sound and serviceable for a good 20 years or more, so the initial outlay is minimal when broken down over such a time period. Less expensive alternatives may not last quite so long, but it is worth weighing up all the options: for example, galvanized-metal drinkers and feeders are expensive but will outlive the less costly plastic ones; on the other hand, plastic is easier to keep clean.

Prices of stock will vary depending on whether you choose commercial hybrids or pure breeds, the latter usually being the more expensive. Point-of-lay stock (see Buying, page 120) will also cost more than youngsters that need bringing on, although you do need to take into account the extra time and food required to raise young birds to maturity.

Provided that you maintain a reasonable common-sense attitude towards the question of hygiene, it is unlikely that your flock will encounter any serious disease problems that might require the services of a veterinary surgeon. That said, there is always the faint possibility that some expensive vet's bills could be encountered at some point.

The daily running costs of keeping chickens are low, and the purchase of pelleted feeds, cereals, grit, vitamins and floor litter will generally amount to very little. You may want to have electric lighting in the henhouse for the winter months or power points from which to run an incubator, so remember to factor in the initial expense of installing these as well as the additional running costs on your electricity bill.

BY-LAWS AND NEIGHBOURS

Before you embark on keeping chickens, check whether any local authority or other regulations are in force that prohibit the practice in your area. If this is the case, either accept that your project is doomed to failure or consider moving house!

Explain your plans to any immediate neighbours. If you do not, you might find yourself involved in a neighbourhood quarrel or even a court case that no amount of eggs offered as a sweetener will rectify.

Rats and other vermin must be kept under control by the use of traps or poison. It is far better to take such preventative steps from the outset rather than wait until any complaints are officially registered; if this happens, you may be legally forced to get rid of your stock, and if you fail to do so you may be prosecuted. Likewise, the owners of a loudly crowing cockerel could find themselves subject to a noise-abatement order.

Depending on the size of your intended set-up, it may be necessary to obtain local authority planning permission if you want to construct a combined chicken shed, veranda and food store or even showing/isolation pens. In most cases it comes down to overall size, whether or not the new building will be attached to an existing one, and its proximity to any neighbour's boundary. Finally, any fencing around your property must be good enough to prevent your birds wandering off your land. In most countries there is a legal requirement for you to protect other people's property from your livestock.

WHY HAVE A COCK BIRD?

Technically, a cockerel is a male bird before it reaches its first adult moult (this usually takes place at around the age of 18 months), after which it becomes known as a cock. Generally, however, the term cockerel is taken to denote a male under 12 months of age. Traditionally, the cock bird holds a time-honoured place in a small flock, protecting his hens, announcing the new day and helping to produce the next generation. But apart

Remember that you don't need to keep a cock bird with your flock unless you intend to breed from it.

from the fact that a cock bird is necessary for breeding purposes and that he looks wonderfully colourful as he struts his stuff, is there any real reason to keep one?

Many people believe that hens will not lay without a cock bird or that they need one to watch over them, but this is untrue. In the absence of a cock, hens will find their own pecking order, with one of the more dominant birds taking on the male's role. There is, however, no doubt that a cock bird enhances the pleasure of keeping chickens, even if there is no real practical reason to have one. After all, there is no scene more pleasant than that of an elegant-looking male drawing his hens' attention to an interesting titbit, then standing proudly over them while they peck excitedly at the treasures he has found.

While a cock bird may look magnificent, on the down side he will lay no eggs, will disturb your neighbours with his early morning crowing, and will eat a lot of food. More seriously, egg production can even go down if hens are stressed and injured by a cock bird's attentions during mating.

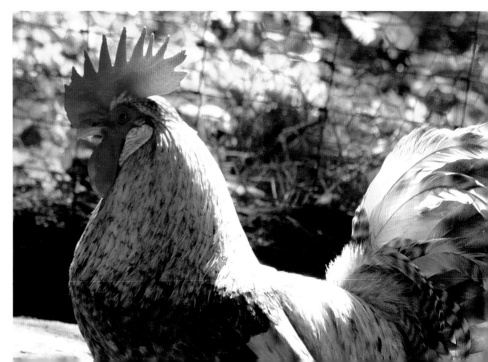

In particular, his spurs and claws can damage their backs and he will also pull out their head feathers while trying to hold on.

OVERCOMING PROBLEMS

Some breeds of chicken are known to be more aggressive than others, so if you decide to keep any of these it will pay not to include a cock bird in the flock unless absolutely necessary. Certain breeds, such as the Ko-Shamo, crow less loudly than others, which may influence your choice. Although it is impossible to stop a cock from crowing altogether, some preventative measures for this and other problems will work with varying degrees of success.

Most complaints against cock birds come down to the fact that they crow at dawn, which can be particularly annoying in the summer months when it gets light very early. A cock needs to raise his head before he can crow, something he will be less inclined to do if the henhouse is a small coop and the perch is fixed high. However, this method works only when he is on the perch and so will simply delay crowing until he is let out.

It is sometimes possible to prevent a cock from crowing too early by creating some sort of shutter system, whereby the chicken house is made totally dark. Make sure that there is still adequate ventilation and fresh air for all the roosting birds. Another option, but one that is time-consuming on a daily basis, is to remove the cockerel each evening and house him in a box or specially constructed cage at the rear of a garden shed, garage or similar outbuilding well away from your house and those of your neighbours.

PREVENTING DAMAGE TO THE HENS

The simplest way to prevent a cock injuring hens is to ensure that the flock is large enough for him to divide his attentions between all the hens rather than concentrating on just a few. In a breeding pen where only one cock and three or four hens of one variety are required, introduce a few extra hens, preferably ones that lay a different-coloured egg so that the correct eggs can be identified for hatching. A ratio of one cock to ten hens seems to alleviate the problem in most breeds. Another option would be to purchase lightweight canvas saddles and attach them to the hens' backs to protect them from damage during mating.

It is also possible to trim the spurs of a cock bird, with either a strong pair of clippers normally used for trimming the claws of dogs, or a hot blade that cauterizes as it cuts (bleeding is always a possibility when an inexperienced person attempts the former method). A traditional method of removing spurs uses a potato that has been baked in the oven for an hour and then cut in half. Protect the cockerel's legs and feet by wrapping them in a towel or rolled sheets of kitchen paper, then push each half of the potato on to a spur. Hold the potato halves in place for five minutes with the aid of a thick gardening glove, then remove. The spurs should then come away with a firm but gentle push and a quick twist.

SELECTING A BREEDING COCK

The cock bird is the most important member of the breeding pen, and if his background is doubtful many of his faults will be transferred to his offspring. He should, therefore, come from quality breeding stock, and be of the type outlined in the breed standards, well muscled and firm to handle. His eye colour should be bold and even, and should surround a perfectly black pupil.

The comb should also conform to the breed requirements and his legs should be clean and strong – without sound legs and feet, he will find it very difficult to mount and mate with his hens. Age is also an important factor: while a cockerel is sexually mature from around six months, it is better to choose a bird that was bred in the previous spring or early summer.

DOUBLE MATING

This method of breeding requires two distinct strains of birds: one to produce exhibition males (this strain is called cock breeders) and the other for exhibition females (pullet breeders). Double mating is usually carried out to create a particular plumage type or colour.

In the partridge Wyandotte, the breed standards require males to have a solid black breast, something that is achieved by crossing indistinctly marked hens with exhibition-standard cock birds. The female chicks produced in this line are not suitable for showing. Exhibition pullets require pencilling, produced by crossing cock birds that display a mixture of brown and black in their breast plumage with exhibition-standard hens. Like the females produced from the cock-breeding strain, the male offspring of pullet breeders are not suitable for exhibition.

If you do decide to keep a breed that requires two strains, your selection of cock birds must obviously reflect this (see Breeding Chickens, page 132). In addition, you will need at least two runs so that the separate lines can be kept apart.

CHICKEN BREEDS

Chickens can be such unique characters that it can be hard not to fall in love with each and every breed. It's worth doing your homework, however, if you want to find your perfect match. Visit shows and exhibitions and spend time talking to breeders. You'll soon discover that knowing something about the various breeds, and being able to distinguish between them, is all part of the fun of keeping chickens.

A poultry exhibition is a wonderful event. No matter whether it be held under canvas as part of a summer agricultural show, in a huge hall rented for the occasion by a national poultry club, or in the committee rooms of your local community centre, which have been lent to the local fancier's society, the first thing you will hear on arriving anywhere nearby is the raucous calling of cockerels anxious to show off. Enter the doors and your nostrils will be assailed by the warm and welcoming smell of fresh wood chippings and, under the by now almost deafening crowing of the male birds, there will be a softer, gentle sound of female poultry crooning and contentedly scratching their floor litter. Look into the pens and not only might you find a freshly laid egg (chickens and bantams do not let the small fact that they are on show deflect them from their daily routine!), but you will also notice that there is a whole new world of feather variations, colour types and body shapes. Some are classically rounded 'puff-balls', whilst others are elongated and arrogant in their pose. With such a plethora of choice, unless you know exactly what breed of bird you are looking for, time spent looking at chickens in any exhibition hall and talking to their breeders is never wasted. Armed with the knowledge you gain and with the help of the information given in this chapter, you will be much better equipped to make that final decision.

Left: A fine example of a single-combed cockerel.

It is, however, not practically possible to cover all the varieties of each breed in a book like this, since in some cases the list would be immense. Also, some breeds and varieties are recognized in the UK and US but not in other parts of the world and so to be specific about absolute colour standards, for example, would leave the book open to criticism. American breed standards classify their birds very differently to the UK and include such things as single comb, clean-legged, Asiatic, English and feather-legged, whilst the show world and standards in the UK use breed types such as heavy, light, soft-feathered, hard-feathered, rare, bantam and true bantam – the difference between 'true' bantams and bantamized versions of large fowl breeds being that the true bantam has no large fowl counterpart. European fanciers tend to speak of 'true' or 'original', and miniatures', but sometimes the true bantam is referred to as a 'natural'. Interestingly, bantams were very much sought after in Victorian times when they were appreciated for their utilitarian qualities, and the fact that they required very little space and could, therefore, be kept in even the smallest of gardens. They were in vogue with the aristocracy, who not only loved their fine appearance, but were also taken with the fact that their eggs contained a much larger yolk-to-egg ratio than could be found in the eggs of any large fowl.

Whether large fowl or bantams are your eventual choice, it is a good idea to use this section in conjunction with any relevant, published Poultry Breed Standards.

Opposite: Attractive though heavily feathered birds may be, their added adornments may cause day-to-day management problems.

American Dominique

Sometimes known simply as the Dominique, this was, for a time, the most common backyard fowl seen in America. The main reasons for the breed's popularity were that it possessed a very hardy constitution and could therefore be kept at little expense by even the poorest of rural families. Heavy plumage protected the birds from the worst of the weather, and their dark and light irregular feather colouring made them practically invisible when scratching in shaded areas – a fact that helped protect them from predators.

The feathering is barred and very similar to that found in the Plymouth Rock (see page 76). The males are generally lighter in marking than the females, but in both sexes, the light bar should always be wider than the dark. There is no colour variation. The comb when viewed from the top should start rounded at the front before flaring out to the sides and finally rounding back to the spike. The modern bird is dual-purpose – providing meat and laying a medium-sized light brown egg.

fact file

Breed Type	Soft-feather. Heavy. Rare
Bantam	Yes in US, but not known in UK
Purpose	Dual
Egg Laying	✔ ✔
Ease of Keeping	✔ ✔ ✔ ✔ ✔
Sociability	✔ ✔ ✔ ✔ ✔

Heavy plumage protected the birds from the worst of the weather.

F

M

Amrock

The Amrock originates from America and is a dual-purpose utility bird that was developed from the Plymouth Rock (see page 76). Nowadays, it is kept mainly by fanciers, who favour it for the fact that it can become very tame with careful handling. It is less likely to go broody than some of the other heavy breeds and it lays a good quantity of large brown eggs.

The Amrock is a handsome bird, deep and rounded in shape. Its colouring is similar to that of the Plymouth Rock and Maran (see page 59), but the barred markings are unique in that the cock bird has white and black bars of equal width, while on the hen the black bars are twice as broad as the white ones. The comb is single, and good specimens have yellow legs, although those of the hens will fade to almost white during periods of intense laying.

In the US, prior to the 1920s, any 'barred coloured' birds were known as Plymouth Rocks and were either light or dark in colour. When the US simplified its exhibition standard to accept only the darker type, the genetics found in the lighter barred birds were incorporated into commercial poultry lines. These were given the name 'Amrock' purely as a marketing tool for importing them into Europe. In recent years European breeders have started showing them and breed clubs have been formed in several countries.

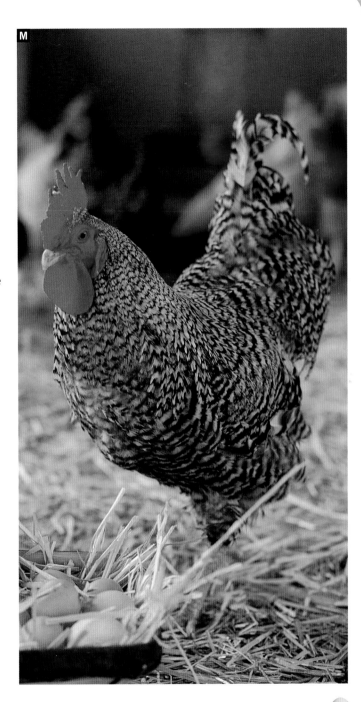

fact file

Breed Type	Soft-feather. Heavy. Rare
Bantam	Yes
Purpose	Dual
Egg Laying	✔✔✔✔
Ease of Keeping	✔✔✔✔
Sociability	✔✔✔✔

Ancona

Anconas are an ideal beginner's bird but can have a tendency towards flightiness. It was once thought that they were derived from the Mottled Leghorn and were, therefore, a member of the Leghorn family (see page 57) rather than a separate breed, but this is now known to be untrue.

Unlike many other modern-day breeds, the Ancona is still pretty much as it was nearly 100 years ago – as evidenced by the following observations in *Fortunes from Eggs*, a book compiled in 1919 by Karswood, who were producers of a very popular poultry spice of the time:

'It is probably the only breed which has the same type for utility and exhibition, wherefore many of the best layers are bred from exhibition winners. The Crystal Palace Laying Competition was won by Anconas [and] at the Missouri Laying Contest in 1914/15, it was found that, on the average of all pens entered, the Anconas laid 300 eggs for every 100lbs of food [compared to], Leghorns 268, Wyandottes 251, Orpingtons 280, Campines 225, Minorcas 203, Rhode Islands 201, Rocks 188, and Langshans 179. This proves that the Ancona is the most profitable over and above cost of food. Also the Ancona eggs were the heaviest of all breeds. Anconas do well intensively; while on free range they pick up half their own food. They are non-sitters, and grow more quickly than any other breed, while the chicks are easy to rear, and offspring comes true to type and colour. Anconas lay well for three seasons, and can be bred from profitably for four. Use rather light-coloured females [for breeding], not too high in tail with legs well apart, deep "bellies" and a fine, falling comb. Mate to a male with a full flow of feather, a high tail, and a light plumage. Avoid males with thick, fleshy combs and coarse lobes.'

Most commonly black or, more accurately, beetle-green in colour, each feather of an Ancona has a white, V-shaped edge to its tip. Chocolate- and blue-coloured birds also occur and similarly have white tips to every feather. Interestingly, the white tips tend to become bigger after each moult. The comb is usually single, but rose-comb varieties are sometimes seen. Anconas are prolific layers of white eggs.

> Unlike many other modern-day breeds, the Ancona is still pretty much as it was nearly 100 years ago.

fact file

Breed Type	Soft-feather. Light
Bantam	Yes
Purpose	Layer
Egg Laying	✔ ✔ ✔ ✔
Ease of Keeping	✔ ✔ ✔ ✔ ✔
Sociability	✔ ✔ ✔

M

The Ancona originated in Italy and was introduced into Britain in 1851, from where it was then exported to other countries, including the United States.

Andalusian

This soft-feathered Mediterranean breed originated in Spain – the bird seen in the UK today was developed from black and white stock imported from there in the mid-1800s. By the mid-20th century the Andalusian had become rare, but it seems to have regained its popularity and it is now possible to see specimens at shows in many countries. In the US, it is known as the Andalusian Blue.

Andalusians display colour variations. Each blue feather is surrounded by black lacing, apart from the sickle feathers of the male and the neck hackles of the female, both of which are black. The comb is single and of medium size: the cock bird's is well serrated and erect, while that of the female normally flops over to one side. The hen is a good layer of white eggs but, like the Ancona, can be nervy and flighty.

· Although a very attractive bird, the Andalusian is definitely not for the beginner, due in no small part to its flighty tendencies and the fact that it prefers free-range to confinement. Breeding exhibition birds can also cause problems as there is a good chance that, even if you mate pedigree blue-coloured parents, as many as 50 per cent of the offspring may mature into birds that carry either black or white splashes, which will preclude them from being shown.

Each blue feather is surrounded by black lacing, apart from the sickle feathers of the male and the neck hackles of the female, both of which are black.

fact file

Breed Type	Soft-feather. Light. Rare
Bantam	Yes
Purpose	Layer
Egg Laying	✔ ✔ ✔
Ease of Keeping	✔ ✔ ✔
Sociability	✔ ✔ ✔

F

Appenzeller

There are three colours of Appenzeller in Great Britain, but the Spitzhauben – silvery-white, with a black tip at the end of each feather – is by far the most popular. The gold-tipped type should have similar markings to the Spitzhauben (with the obvious difference being in the silver/white and gold 'base' colouring), and there are also blacks, but these are the least common of the three colours.

The name 'Spitzhauben' means 'pointed hat or bonnet' – a possible reference to the traditional Swiss bonnet of its country of origin – and the breed is sometimes also referred to as the 'Pointed Hood'. It has a crest and V-shaped comb. The Appenzeller Barthuhner (known in America and elsewhere as the 'Bearded Hen') is a more powerfully built bird which lacks the head crest of the Spitzhauben – from which it also differs in that it has a rose comb with a backward-pointing spike, a full beard and white ear lobes.

In the UK the Appenzeller is a relatively new breed, having been introduced in 1982 by Her Grace the Duchess of Devonshire (now the Dowager Duchess). Despite having its own breed club, it has not really gained many enthusiasts, not least because of its propensity to flightiness, its loathing of confinement and its preference for roosting in trees rather than on the perches of the chicken house. At the time of writing, there are no British standardized Appenzeller bantams, although the bantam is officially recognized in the US.

fact file

Breed Type	Soft-feather. Light. Rare
Bantam	Yes in US; not known in UK (see text)
Purpose	Layer
Egg Laying	✔✔✔
Ease of Keeping	✔✔✔
Sociability	✔✔✔

Araucana

In most countries, two types of Araucana are seen: tailed and rumpless. The rumpless variety is more akin to the original breed first kept many years ago by the Arauca Indians of Chile, but today it is the slightly rarer of the two.

A further defining feature of the breed is the tufts of feathers sticking out of the birds' ears; known as ear rings or ear tufts, ideally these should slant backwards. The tailed variety also has feathered head crests and beards, and both types have a pea comb that is irregular in shape.

The breed contains at least 11 colour varieties. The most popular of these is lavender, but it also comes in partridge, silver-blue partridge, yellow partridge, fawn, wheaten, white, black and cuckoo. Another unusual characteristic is the fact that the hens lay blue-green eggs (US standards describe the colour as turquoise). Like the Andalusian, the Araucana is a good layer and, being hardy, makes an interesting choice for the novice chicken keeper.

Neither a spelling mistake nor the result of a typesetter getting a finger stuck on the 'a' of the keyboard is the Ameraucana, a breed that is sometimes confused with the Araucana. There are several similarities in that both lay a blue-coloured egg, have pea combs, red ear lobes and, when *not* of the rumpless type of Araucana, have ear tufts, head crest and muffs.

The Araucana's distinctive plumage makes it an attractive choice for the keeper who also wants plenty of eggs.

Aseel

Also sometimes spelled Asil, the Aseel was instrumental in the make-up of the Indian Game or Cornish (see page 51). Originating in India, it is reckoned to be one of the purest pure breeds in the world and is certainly one of the oldest, having been around for at least 2,000 years. It has broad, prominent shoulders, wide hips, a narrow stern and a pea-shaped comb.

The breed should have a moderately low carriage and powerful limbs. The small pea comb, coupled with practically no wattles is a very desirable point as far as the show bird is concerned and a legacy of the days when they were used for fighting: the minimum amount of head 'furniture' would have given little for an opponent to use as a point of contact. Small combs and wattles also negated the need for the combs to be dubbed.

The breed's fighting ancestry is still apparent in today's birds and both males and females can be highly aggressive: cock birds will frequently fight to the death.

Although known in the US, Aseels are perhaps slightly more common in the UK and Europe, where they are bred in various colours of plumage, including black, white, duckwing, dark red, light red, grey and pile. The hens lay tinted eggs and will go broody. Like the Indian Game, they can be difficult to breed successfully, as the cock bird sometimes cannot physically manage the mating process due to his top-heavy build.

fact file

Breed Type	Hard-feather. Heavy. Rare
Bantam	Yes
Purpose	Exhibition
Egg Laying	✔
Ease of Keeping	✔✔
Sociability	✔

The breed's fighting ancestry is still apparent in today's birds and both males and females can be highly aggressive.

M

Australorp

Australorps are easy to keep and become very tame, so they make good pets for children.

fact file

Breed Type	Soft-feather. Heavy
Bantam	Yes
Purpose	Dual
Egg Laying	✔✔✔✔✔
Ease of Keeping	✔✔✔✔✔
Sociability	✔✔✔✔✔

The name of this breed is an abbreviation of 'Australian Black Orpington'. It resulted from stock birds imported to the UK from Australia around the 1920s being crossed with the Black Orpington, which was developed by William Cook in the late 1800s (see page 72). The Australorp can, therefore, be said to be both British and Australian in origin.

This soft-feathered bird is heavy and traditionally black, although white- and blue-laced variations do occur. The comb is single, serrated and of medium size, and the eggs are tinted towards brown in colour.

Australorps are easy to keep and become very tame, so they make good pets for children. Unlike some other breeds, they are not aggressive towards one another – even young cockerels being reared as part of a breeding programme can normally be kept together without mishap.

The Australorp is an active breed and, although the birds will happily live in runs, they do enjoy being allowed to free-range. Not being good fliers, they don't need particularly high fencing around a run in order to contain them. For a heavy breed, they are amazingly productive and their egg-laying performance attracted world attention when in 1922-23 a small flock of six hens set a world record of 1,857 eggs at an average of 309.5 eggs per hen over 365 consecutive days in a trial held in Australia – and that without any of the artificial lighting normally given to commercial breeds during the winter months. A few years later,

a new record was famously set when a single hen laid 364 eggs in 365 days.

With such a reputation for domesticity, it is not surprising that Australorps are particularly popular in their 'home' country and that they are well exhibited at shows all around Australia. Australia's standard for Australorps is very similar to the British Poultry Standard, but it is interesting to note that although the white and blue-laced types are known, colours other than black (in reality, a beautiful, iridescent, blue-green sheen) have never really taken off in Australia; even though blue-laced bantams are standardized, they are seldom seen on the show circuit.

Australorp fanciers everywhere argue regarding the merits of bantams over large fowl, with a great number coming down in favour of the bantam type. Reasons given include the facts that they lay a good-sized egg (although they will not lay as many – approximately 170 per season as opposed to as many as 260 a year by the standard fowl) and they are cheaper to purchase initially. Some also say that the bantams are hardier than the large fowl, but there is no proven justification for this statement. No matter which size you choose, all types of Australorps make an ideal bird for the beginner.

Barnevelder

Imported from the Netherlands to the UK and the US in the early 1920s, the Barnevelder is thought to have Cochin, Langshan and Brahma genes in its make-up (see pages 39, 41 and 36), a mixture that has resulted in a large range of colours. Nowadays, the only colours accepted as being in accordance with most countries' breed standards are black, white, double-laced and blue double-laced, although in Germany blue and brown are also permitted on the show bench.

By far the most popular, and arguably the most attractive, is the double-laced version, which has dark hackle feathers with a bottle-green sheen, although technically, in Barnevelders, only the females are double-laced and the males in their final plumage are what is more correctly known as melanistic black-breasted reds. The edge markings of the cockerel's feathers appear iridescent green in the light and its tail feathers are black, but on the hen bird the lacing continues right into the tail. All posess a single comb.

Barnvelders are good-tempered and sociable, making them an ideal choice for a beginner.

fact file

Breed Type	Soft-feather. Heavy
Bantam	Yes
Purpose	Dual
Egg Laying	✔ ✔ ✔
Ease of Keeping	✔ ✔ ✔ ✔ ✔
Sociability	✔ ✔ ✔ ✔ ✔

The iridescent green sheen on the male's feather makes this a particularly attractive bird.

The double-laced large fowl Barnevelder originated at the same time as the breed as a whole was being developed, but the bantam type was not recognized until the 1930s: it was bred in Germany by crossing small large fowl with bantam Rhode Island Reds and Gold-laced Wyandottes (see pages 80 and 102) amongst others. The bantam breed of Barnevelder soon gained worldwide recognition and was exported to many countries because of its ability to lay approximately 180-200 large brown eggs per year. There has been, unfortunately, less emphasis placed on the egg colour in recent years and so, if it is the traditional dark-coloured egg that attracts, it will pay to be very careful when sourcing a particular strain of bantam from a breeder.

As far as the large fowl is concerned, any of the colour varieties would make a good choice for the first-time chicken fancier, as they are all easy to keep – either free-range or in a grass run – and, like the Australorp, have an easy-going and placid temperament. In addition, the hens lay a good number of dark brown eggs through the year. Because Barnevelders are not generally well represented on the show bench, this is a good breed for the novice exhibitor to choose, as there is a fair chance of winning some classes during the first season of showing.

Barbu d'Anvers, Barbu d'Uccle and Barbu de Watermael

Collectively known as Belgium bantams, these are all close cousins and are true bantams. Unsurprisingly, given their collective name, they are thought to originate from Belgium, but their country of origin is sometimes given as the Netherlands – a result of their also being found in the Flemish-speaking part of Holland lying near the Belgium border.

Belgian bantams were first brought to the UK in 1911 after a display was staged at the Crystal Palace. In Australia, however, the Belgium bantam breeds are relative newcomers, having been known there only for a matter of 30 years or so.

Irrespective of type, all have the same colour variations and lay cream-coloured eggs. All are bearded and muffed, but the Barbu d'Anvers has clean legs and a rose comb, whilst the d'Uccle has a single comb and heavily feathered legs and feet. The Barbu de Watermael has a rose comb and a crest. The feathered legs of the d'Uccle mean that these birds generally require a little more attention to their welfare than some other breeds, especially when considering them for showing and exhibition purposes, but much can be done by keeping them in covered runs with the floors littered with silver sand.

The d'Anvers is also known in some places as the Belgium d'Anvers or even the Antwerp Belgium. Rumpless versions are given the name Barbu du Grubbe while rumpless d'Uccles are called Barbu d'Everberg. In America, Belgium bantams are known and prefixed as 'Bearded' rather than 'Barbu'. All are commonly seen as the 'mille fleur' type (the word means 'thousand flowers') which was the original colouring and certainly the first to be imported to the US; other colours include black, blue, mottled, porcelain, quail, self-blue and white.

As ornamental pets, Belgium bantams make attractive additions to the garden, but it is as well not to make the mistake of thinking that, because of their diminutive size, they will merely potter about looking pretty – they are very active and their light weight and small size make it possible for them to fly into next door's garden very easily. Because of this it is perhaps advisable to keep them in a closed pen in any situations where neighbours may be a problem. The birds will quickly become tame whether confined or given free-range, and their colour variations and ornamental stature make them well worth considering.

fact file

Breed Type	True bantam
Bantam	Bantam only
Purpose	Ornamental/exhibition
Egg Laying	✔
Ease of Keeping	✔✔✔
Sociability	✔✔✔✔

It is as well not to make the mistake of thinking that they will merely potter about looking pretty.

F

Booted

Sometimes also known as Sabelpoots, Booted bantams are similar in appearance to the Belgium bantams (see previous page) but the Booted doesn't have the beard and muffs, its main characteristic being its quite substantial foot feathering. To maintain these feathers in good condition, great care must be taken and many showing enthusiasts keep the birds either indoors or in enclosed runs with the floors littered with silver sand.

The breed's foot feathering, together with the fact that it is more able to flutter than to fly, means that any perches incorporated into the housing will need to be somewhat wider and lower than is the case for most other true bantam types. Even though they can be difficult to keep, Booteds are nevertheless very popular throughout Europe, possibly because they are less likely to fly than are the Belgium breeds and, because of their heavy feathering, unable to cause damage to the flower beds if kept as pets and given free-range of the garden.

Some books give this breed's country of origin as Belgium but, given the fact that they are sometimes known as Dutch Booted bantams, it is more likely that they came initially from Holland. The UK bloodlines were nearly all lost in the early part of the 20th century and it is only through the efforts of dedicated breeders that the Booted exists today – albeit as a rare breed.

fact file

Breed Type	Soft-feather. True bantam. Rare
Bantam	Bantam only
Purpose	Ornamental/exhibition
Egg Laying	✔
Ease of Keeping	✔ ✔ ✔
Sociability	✔ ✔ ✔ ✔

F

Brabanter

A crested and muffled breed with a V-shaped horn comb, the Brabanter is thought to be Dutch in origin. It is considered more of an ornamental breed than a practical one, and although the hens lay decent-sized white eggs, these tend to be produced in the spring and early summer only.

As an ornamental bird, the Brabanter is appealing for its shape, size and colour rather than its egg-laying or meat-producing characteristics. It is found in black, white, spangled, buff, mahogany, cuckoo and lavender colours, and its crest grows vertically, unlike that of most other crested breeds, in which the feathers grow horizontally towards the back of the head. In addition, both male and female have an impressive full throat-beard set between the ear muffs. Of the ornamental breeds, the Brabanter is probably one of the easiest to care for as the birds are good foragers and fairly docile. Since they have small wattles and combs, they are not prone to frostbite, a point that may be worth noting if you are considering keeping them in extreme parts of the US (where they are currently rare). On the downside, they are good at flying and climbing, which may cause problems if you intend to keep them free-range: fortunately, however, the breed adapts well to confinement.

The Brabanter is quite closely related to the Owl Beard (see page 74, another ornamental breed, possessing a beautiful upright carriage and sloping, long backline), the only differences being in the crest and the shape of the beard.

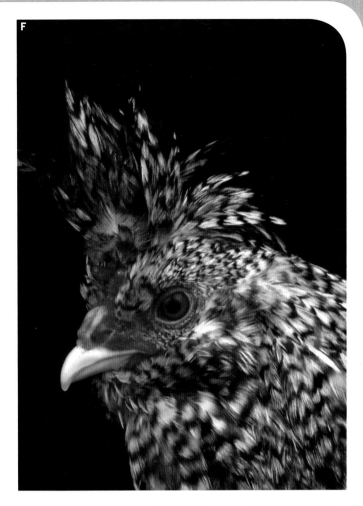

> The Brabanter is appealing for its shape, size and colour rather than its egg-laying or meat-producing characteristics.

fact file

Breed Type	Soft-feather. Light. Rare
Bantam	Yes
Purpose	Ornamental/exhibition
Egg Laying	✔
Ease of Keeping	✔ ✔ ✔
Sociability	✔ ✔ ✔ ✔

Brahma

While the most commonly seen Brahma in many countries is the multiple-pencilled partridge or triple-laced variety, other colours include light, dark, Columbian, buff Columbian, cuckoo, birchen, white and gold. Gold cock birds have glossy black breasts, legs and tails, while the hens have delicate black markings on a gold ground colour.

The Brahma originated in India and reached the UK in the mid-1850s, but was recorded in the US roughly a decade earlier. When the birds were first introduced to the UK, they were known as Brahma Poutras and were a great favourite of Prince Albert.

Sometimes known as the 'king of chickens' because of their great size – the cocks can weigh 5kg (11lb) – Brahmas are profusely feathered, staid and matronly in appearance. The feet and legs are heavily feathered and the comb is triple or pea. For a heavy breed, the hens lay a surprising number of light brown eggs each year but, as is often the case with large varieties, these are relatively small. Unlike many types of poultry, where one cock bird can easily fertilize six or more hens, Brahmas are best kept as trios if you intend breeding them.

fact file

Breed Type	Soft-feather. Heavy
Bantam	Yes
Purpose	Dual
Egg Laying	✔✔
Ease of Keeping	✔✔✔✔
Sociability	✔✔✔✔✔

M

Campines

Like the Light Sussex (see page 94), Rhode Island Red (page 80) and a few other well-known breeds, the Gold Campine has been used over the years to produce a reliable sex-linked breed commercially known as the Cambar. As a pure breed, the Campine is seen in two colours: gold and silver. The difference in colouring is most clearly seen in the head and neck feathers, although the gold and silver do permeate through the body feathers, which are predominantly barred with beetle-green.

Although they are good layers, Campines are considered more as ornamental in North America and are few enough in number to be classified as 'rare' in the UK – and this despite the fact that they were sufficiently popular to have had their own breed society in the late 19th century. One of their most notable characteristics is the cock bird's very 'hen-like' shape: it lacks the long, curved tail, saddle and hackle feathers normally seen in males and, at a quick glance, could be mistaken for a hen.

Campines are generally thought to have originated in the regions around Antwerp in Belgium; nevertheless, Carol Ekarius's recent book *Pocketful of Poultry* suggests that they are descended from the Egyptian Fayoumi, a small breed which has been reared along the banks of the River Nile for many centuries. Mention must also be made here of another Belgian breed, the Braekel, which is similar to the Campine in all respects except size: the Braekel is the larger of the two.

F

fact file

Breed Type	Soft-feather. Light. Rare
Bantam	Yes
Purpose	Layer
Egg Laying	✔✔✔
Ease of Keeping	✔✔✔✔
Sociability	✔✔✔✔✔

Catalana

fact file

Breed Type	Soft-feather. Light. Rare
Bantam	Yes
Purpose	Dual
Egg Laying	✔✔✔
Ease of Keeping	✔✔✔✔
Sociability	✔✔✔✔

It should be considered only by those with plenty of land over which it can be given free-range

The Buff Catalana is a medium-sized bird and, like the American Dominique (see page 20), is noted for its hardiness. Unfortunately for any would-be fanciers, it is also like the Appenzeller (see page 25) in that it does not take kindly to confinement and should be considered only by those with plenty of land over which it can be given free-range.

The breed is not well known in the UK or US, but it is widely distributed through South America – a fact that is perhaps unsurprising due to its having originated in Spain (in the district of Prat near Barcelona) and no doubt having been taken by emigrant Spaniards many centuries ago. Unlike many of the Mediterranean breeds, Catalanas are of medium size and therefore come closer to being dual-purpose than most others. Somewhat plain to look at, they are buff with a greenish-black tail and possess a single comb and large, bright red wattles.

As might be expected of a Mediterranean species, these are good layers of cream/tinted eggs and do not often go broody. Despite not being well known in the US, Catalanas are included in the American Standard of Perfection – the official classification of all recognized breeds of poultry. Buff Catalanas are also known as Catalanas del Prat Leonada.

Cochin

The earliest examples of this breed began to arrive in the UK from the Far East in around 1840. At first the breed assumed different names and it is still sometimes erroneously referred to as the Pekin. In the UK, there is a true bantam known as the Pekin (see page 75) and also a bantam version of the Cochin. In other parts of Europe and in the US, however, only the large Cochin and Cochin bantam are known.

These are big birds and could easily compete with the Brahma for the title of 'king of chickens'. The hocks are covered in feathers that, unusually, curl around the joints, and the legs and feet are also feathered. The comb is small, single and straight, and in particularly good specimens is well serrated. The eggs are light brown and, for a heavy breed, quite numerous. There are several plumage varieties, the most common being black, white and lavender.

As with all feather-legged breeds, Cochins require special attention, and show birds should always be kept on clean, dry flooring. Nevertheless, they do make a good choice for the novice, especially anyone looking for a breed that will reliably go broody.

fact file

Breed Type	Soft-feather. Heavy
Bantam	Yes
Purpose	Exhibition
Egg Laying	✔✔
Ease of Keeping	✔✔✔
Sociability	✔✔✔✔✔

Crève-Coeur

The Crève-Coeur is one of France's oldest breeds and, like the Bresse, which is the best known of all French chickens, was originally bred as a table bird. The modern-day type is larger than its ancestors, owing to the fact that it was crossed with the Dorking from the UK (see page 43) sometime during the 1800s.

The Crève-Coeur's head is topped by a full crest, in front of which can be seen a horn comb; like that of the La Flèche (another ancient French breed, see page 55), this forms a pronounced V-shape. Although it is essentially a table bird, the Crève-Coeur lays a substantial number of white eggs. As it can also become hand-tame reasonably quickly, it is a good choice for those who want a breed that is a little out of the ordinary. However, its crest can attract lice, so it is important to check it regularly and treat any infestation at the first signs.

Another potential problem is that it is not unknown for a Crève-Coeur's crest to freeze up in cold weather. Also, the feathering in general may make it unsuitable for locations where inclement weather is the norm rather than the exception.

Spelt as 'Crève-Coeur' in most countries, general references to the breed in the US have the spelling as 'crevecoeur'. In its original French form the name means 'broken heart', presumably, an allusion to the shape of the comb.

fact file

Breed Type	Soft-feather. Heavy. Rare
Bantam	Yes
Purpose	Dual
Egg Laying	✔ ✔ ✔
Ease of Keeping	✔ ✔ ✔
Sociability	✔ ✔ ✔ ✔

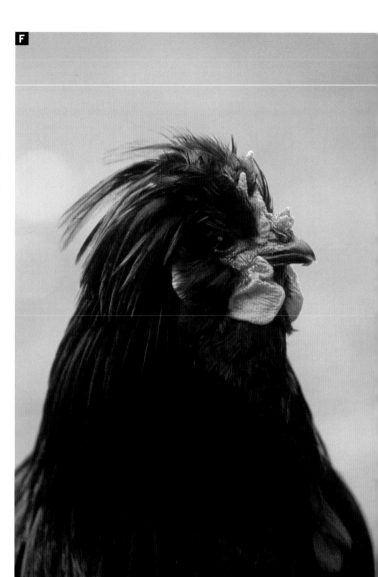

F

Croad Langshan

Originating in China, this breed was imported to the UK by a Major Croad, after whom it is named. It is sometimes also called the Black Croad Langshan, as a white strain does exist but this is rarely seen and the black colouring predominates. In Germany, breeders crossed Major Croad's birds with Minorcas and Plymouth Rocks (see pages 61 and 76) in an effort to achieve a more commercial egg-layer, and in doing so developed the separate Langshan breed.

The breed standards classify both Croad Langshans and Langshans as heavy and soft-feathered. Both have a single comb that should have five points, but while the Langshan has clean legs, the Croad Langshan has sparsely feathered ones. The hens lay well and their eggs are light brown to cream in colour.

The original Black Langshans were imported to North America in 1878 and admitted to the standard in 1883, White Langshans being admitted some ten years later. There are now three varieties of Langshans in the US standard: black, white and blue (the last having been accepted as recently as 1987), but Croad Langhans, as a distinct type, are still not recognized in the US, despite the fact that, along with the Brahma, they were instrumental in the creation of of the Jersey Giant (see page 54).

Although it can be tricky to source, the Croad Langshan is one of the easier breeds that the newcomer to poultry-keeping might consider. Some thought must, however, be given to the fact that their feathered legs might be a disadvantage in wet, clay-type soil conditions.

fact file

Breed Type	Soft-feather. Heavy
Bantam	Yes
Purpose	Layer
Egg Laying	✔✔✔
Ease of Keeping	✔✔✔✔
Sociability	✔✔✔✔✔

Derbyshire Redcap

Although it is also known simply as the Redcap, this breed retains a close association with the English county where it originated. It was originally produced as a dual-purpose bird, but enthusiasts today prize it for its rarity and for its ability to lay numerous white eggs.

The feathers of the male's neck, hackle and saddle each have a red quill with beetle-green webbing, finely fringed and tipped with black. The back is red, while the breast and under-parts are black. The hen is more nut-brown in colour, with a black half-moon spangle at the end of each feather. The tail is black in both sexes. The comb is an accentuated rose in shape and has a long, straight leader.

Redcaps are hardy birds and are at their happiest free-ranging: they are terrific foragers and good fliers, as a result of which they require a lot of space. They are long-lived and will continue to lay plenty of eggs for several seasons but, on the downside, they are not very good at sitting their own eggs.

The Derbyshire Redcap Club in the UK has suffered mixed fortunes over the years: it was very active until the late 1930s, then fell into decline. In June 1976 local breeders re-formed the club, and today it is thriving, due no doubt to the resurgence of interest in the breed.

fact file

Breed Type	Soft-feather. Light
Bantam	Yes
Purpose	Dual
Egg Laying	✔✔✔
Ease of Keeping	✔✔✔✔
Sociability	✔✔✔✔

The comb is an accentuated rose in shape and has a long, straight leader.

M

Dorking

Possibly the oldest of English breeds, the Dorking was certainly around at the time of the Roman invasion of Britain in AD 43. Many other breeds – including the Crève-Coeur (see page 40) – were crossed with the Dorking during the 19th century to improve their suitability for the table.

Interestingly, the colour variations of silver-grey, red, white and cuckoo dictate the comb type of the Dorking: examples of the silver-grey and red should bear a single comb, while the white and cuckoo must possess a rose comb. Other peculiarities of the breed are its five toes and the pronounced boat shape of its body. The hens lay cream-coloured eggs but generally only in season (spring and summer), so cannot be relied upon for a year-round supply.

Some strains of the breed are known to suffer from poor feathering and benefit from being rained on, which will keep the feathers in 'tight' condition. If you are intending to show your birds, however, this will not be possible, as the sun and rain will cause any light-coloured or white feathers to turn brassy.

As a utility bird, the Dorking can still hold its own amongst the more commercial breeds: try crossing it with an Indian Game or Cornish (see page 51), for example, in order to produce an excellent-tasting chicken.

fact file

Breed Type	Soft-feather. Heavy
Bantam	Yes
Purpose	Dual
Egg Laying	✔ ✔
Ease of Keeping	✔ ✔ ✔ ✔ ✔
Sociability	✔ ✔ ✔ ✔ ✔

"Some strains of the breed are known to suffer from poor feathering and benefit from being rained on."

F

Dutch

fact file

Breed Type	True bantam
Bantam	Bantam only
Purpose	Ornamental/exhibition
Egg Laying	✔✔
Ease of Keeping	✔✔✔✔
Sociability	✔✔✔✔

This is another true bantam and is sometimes known as Old Dutch, although the purists say that this is an incorrect title. The Dutch Poultry Club of Holland first standardized the Dutch bantam in 1906, and today their standard recognizes over 20 varieties.

Bred from several regional sources, the breed was at first simply known as the Dutch bantam, and very quickly became popular throughout Holland. It it is still one of the most common breeds in its native country, is well-liked in Great Britain and Germany, and is currently gaining status in both South Africa and the United States. The first colours were the gold and silver partridge, white and black, with others quickly following, and there are now 13 recognized colours in the UK and at least ten in America.

Despite being brought into Britain only in the early 1970s, the Dutch bantam has created for itself a reputation for being an ideal breed for keeping in a small area or a confined space – although, like most breeds, the more space you can give it, the better it will fare. They are hardy birds, but they also thrive on loving care and attention and there-fore make ideal family pets (seek out strains that are known to be particularly placid), as well as laying a reasonable number of small cream eggs and, surprisingly for such small birds, being excellent and reliable broodies.

In America, there is the opportunity for some confusion between the name 'Dutch' and the 'Bassett' as, although in other parts of the world there is a breed known as the Bassette (with the 'e'), in the US Bassett also refers to a Bob Bassett of Florida who served as secretary-treasurer of the American Dutch Bantam Society and so 'Bassett bantam' is the nickname of the Dutch bantam.

The proper Bassette (the name means 'small appearance') is quite rare anywhere but Europe and is most frequently seen in Holland and the southern French-speaking part of Belgium. Bassettes are very lively, are constantly on the move and quickly become bored, so require large pens and distractions in the form of titbits. Experienced Bassette breeders claim that young cockerels have to be separated from their brood sisters far earlier than other ornamental bantams because they are very quick-maturing and will attempt to mate with the females at an extremely young age.

Faverolles

The Faverolles originates in northern France and was bred for both eggs and the table. Nowadays, three distinct types are found: the original French, the German and the British. The breed reached the UK in 1886, since when developments in breed standards have meant that the British type carries its tail higher than its German and French cousins.

Despite their heavy appearance Faverolles are good layers, producing creamy eggs that sometimes tend towards light brown. The breed has five toes like the Dorking (with which it was crossed to produce a larger bird, see page 43), lightly feathered legs, a beard and side muffs. The original colour was known as salmon, and in some countries this is still the only colour permitted on the show bench. In Germany and the UK, however, white, black, ermine, blue and Columbian are accepted, while in France the cuckoo colouring is quite common.

Colour variations in the US are restricted to salmon or white. In Australia, the colours permitted are more like those in the UK, including birchen as well as black, buff, cuckoo, ermine, white and blue-laced. The blue-laced is particularly attractive, with each feather being edged with a darker shade of the same colour. Attempting to breed some of the more complicated colour variations can be difficult and many require separate pens in order to produce cocks and hens suitable for the show bench.

Friesian

Classified as a rare breed in the UK by the Rare Poultry Society, the Friesian is a hardy, light, egg-laying bird that originated in the northern parts of the Netherlands. Even the large variety is very small, to the extent that it is not unknown for standard birds to be entered in bantam classes by mistake.

Friesians produce white eggs, are upright and bold looking, and have quite pronounced white lobes. Their legs are slate-blue and the comb is single. Colours can include black, white, red and cuckoo, with the five further varieties – gold, silver, red, lemon and chamois – all being described as pencilled (having numerous small black dots on the feathers of the breast, back and wing shoulders). Although the shape of pencilling varies from breed to breed, in the Friesian the black spots are tiny and teardrop-shaped. As Friesians are light, they are good fliers and it may therefore prove necessary to keep them in an enclosed run to prevent them from roosting in the trees of a neighbour's garden. As with any of the other light, flighty breeds, it would be possible, of course, to clip the flight feathers of one wing, but the bird could not then be shown until the feathers had regrown during the following autumn's moult.

fact file

Breed Type	Soft-feather. Light. Rare
Bantam	Yes
Purpose	Layer
Egg Laying	✔✔✔
Ease of Keeping	✔✔✔
Sociability	✔✔✔✔

As Friesians are light, they are good fliers and it may therefore prove necessary to keep them in an enclosed run.

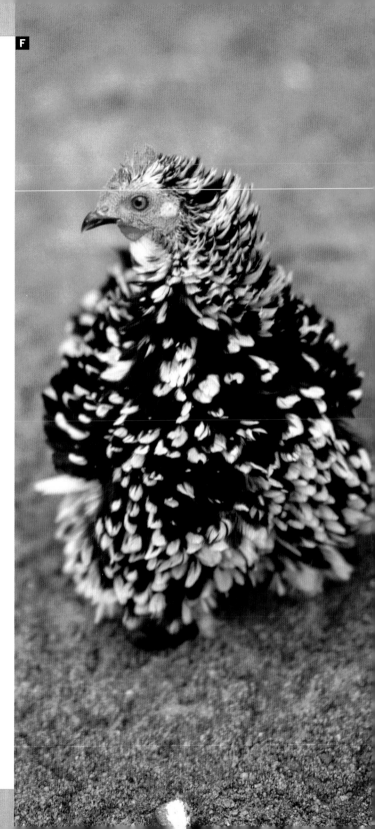

Frizzle

The Frizzle is Asiatic in origin and is generally thought of as an exhibition breed, the bantam variety having always been more popular than its larger relation. It gets its name from its strange feathering, which should curl towards the head and be as tight and even as possible.

In some countries, the word 'frizzle' simply denotes a feather type, but in the UK it is a recognized breed. If you breed from the same strain year after year there is a danger that you will end up with birds that have sparse and weak feathering. To avoid this always use a fresh, well-frizzled male to mate with your pullets.

Frizzles come in a range of colours, the most common being black, white, buff, blue and silver-grey. Other permitted but less common colours include Columbian, duckwing, black-red, brown-red, cuckoo, pile and spangle. Even though it is an exhibition breed, the Frizzle lays reasonably well (the eggs are cream or tinted in colour). Anyone considering keeping these birds should, however, be aware that their feather formation leaves them unable to cope with particularly wet weather, so they are better housed indoors.

One thing that exhibitors of Frizzles see as a constant battle is the attitude of people who will not treat the breed with the seriousness they feel it deserves, some regarding Frizzles as hybrids rather than a breed in its own right.

NB: As well as being the name of this breed, 'frizzled' can mean a fault in other poultry types and is normally associated with poor plumage.

fact file

Breed Type	Soft-feather. Heavy
Bantam	Yes
Purpose	Exhibition
Egg Laying	✔ ✔ ✔
Ease of Keeping	✔ ✔ ✔
Sociability	✔ ✔ ✔ ✔

Hamburg

Although Hamburgs originated in northern Europe, both spangled and black varieties have been bred for more than 300 years in northern England, where until the mid-19th century they were known as 'Pheasants' and 'Mooneys'. In the UK, Hamburgs are also known as Hollands, whereas in the US Holland is a separate breed. Hamburgs are smart and elegant, with a rose comb and white ear lobes. They lay a reasonable number of white eggs, but are known to be aggressive towards each other if not given sufficient space in the henhouse and in the run.

Silver Spangled Hamburgs are white with a large black spot at the end of each feather, while Gold Spangled Hamburgs are a rich mahogany colour, with the same black spots. These attractive birds will enhance any garden, although the spangled varieties may become discoloured or brassy if exposed to too much sun. These feathers will eventually moult out, but any suspicion of brassiness would put paid to a bird's show prospects. Pencilled Gold and Silver Hamburgs are created from separate cock-breeder and pullet-breeder strains through a process known as double mating (see page 17) and is usually carried out to create a particular plumage type or colour. Another example in which double mating is necessary is the Partridge type of the Wyandotte breed (see page 102).

fact file

Breed Type	Soft-feather. Light
Bantam	Yes
Purpose	Exhibition/layer
Egg Laying	✔✔✔
Ease of Keeping	✔✔✔
Sociability	✔✔✔

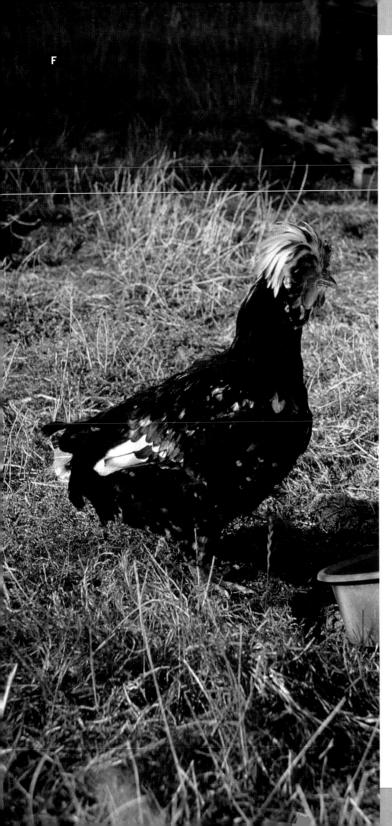

Houdan

Introduced to the UK in the mid-1850s from France, where they were first bred as a table bird, Houdans are now used mainly for exhibition purposes, although they are also reasonably good layers of white eggs. Its colours are limited, and although white and lavender are seen in some countries, the black mottled variety is the one most often encountered on the show bench.

The Houdan is heavily crested, probably due to the fact that, generations ago, Crève-Coeur blood (see page 40) was introduced to the breed. The distinctive crest, the complete muffling surrounding the face and the leaf-shaped comb (also called a butterfly) together make the Houdan a rarity in the poultry world. To describe the comb further, it is perhaps easier to say that it is in the shape of an oak leaf and looks like two single combs connected only at the point above the beak. As with other crested breeds, the Houdan needs a little extra care in terms of its housing and is best kept indoors or in a covered run. Like the Dorking and Faverolles (see pages 43 and 46), it is unusual in having five toes.

fact file

Breed Type	Soft-feather. Heavy. Rare
Bantam	Yes
Purpose	Exhibition/dual
Egg Laying	✔✔
Ease of Keeping	✔✔✔
Sociability	✔✔✔✔

Indian Game

Originally bred in Cornwall for cock fighting, Indian Game birds were subsequently developed for the table through the introduction of blood from foreign breeds brought into the UK by sailors travelling back from Asia. Because of their origins, they are also known as Cornish Game birds, especially in the US.

Indian Game possess pea combs and red ear lobes. Only three colours are recognized in the UK: dark (chestnut double-laced with greenish black), jubilee (chestnut laced with white) and blue. Elsewhere, a white laced with chestnut is acceptable.

They are chunky in appearance and, with their widespread legs, look a little like a bulldog when viewed from the front. From above and behind, they are wedge-shaped, with the body tapering down to a short but thick-set tail. It is sometimes difficult to breed them, as the cock bird often cannot physically manage the mating process owing to his top-heavy build. The male is unusual in being almost monogamous, and so any breeding pens should contain only pairs or trios.

Despite being bred originally as fighters, nowadays Indian Game birds are good-tempered and can become tame, but they are not good layers, producing only a few cream or light brown eggs in the late spring/early summer months.

> They are chunky in appearance and, with their widespread legs, look a little like a bulldog when viewed from the front.

fact file

Breed Type	Hard-feather. Heavy
Bantam	Yes
Purpose	Table
Egg Laying	✔
Ease of Keeping	✔ ✔ ✔
Sociability	✔ ✔

Ixworth

Compared with many other breeds of poultry that have been around for several generations, the Ixworth is a relative newcomer and was first produced in the period between the two World Wars.

Taking its name from its Suffolk village of origin, the breed was developed by Reginald Appleyard, who set out to produce a first-class, quick-maturing table breed that could also lay a reasonable number of good-sized eggs. He used several breeds to create the Ixworth, including White Old English Game,

Indian Game, White Sussex, White Orpington and White Minorca – all in an effort to produce the fine white skin required by the English table market.

Ixworths are quite deceptive to look at, being a great deal heavier and more solid than they appear. They are active, hardy birds which do well on free-range. There is only the one variety, white, which has pinkish-white legs and beak, and all-white plumage. The comb is always of the pea type and the eyes are orange to red in colour.

The breed very nearly died out some 40 years ago, but thankfully a few dedicated and enthusiastic fanciers were on hand to keep things going. The future for this truly English breed is now far more positive than it once was, but it is still very much a rarity and so, should you be tempted into choosing Ixworths, make every effort to ensure that you are being offered stock with provenance and pedigree rather than birds that are a hybrid approximation of those developed by Appleyard.

Ixworths are quite deceptive to look at, being a great deal heavier and more solid than they appear.

fact file

Breed Type	Soft-feather. Heavy. Rare
Bantam	Yes
Purpose	Dual
Egg Laying	✔✔✔
Ease of Keeping	✔✔✔✔
Sociability	✔✔✔✔

fact file

Breed Type	Soft-feather. True bantam
Bantam	Bantam only
Purpose	Ornamental/exhibition
Egg Laying	✔
Ease of Keeping	✔ ✔
Sociability	✔ ✔ ✔

Japanese

This is another true bantam breed and perhaps its most noticeable features are the very distinctive, long, upright tail carriage of the cock bird and the short legs of both sexes – so short, in fact, that they sometimes cannot be seen because they are covered by the thigh feathers.

The breed has been known for many years in Japan, but is a comparative newcomer to Europe where it first appeared in the mid 1800s. Originally, it would have been unusual to see anything but the feather type observed on most 'normal' poultry breeds, but since the 1960s two other feather variations have become increasingly popular amongst Japanese bantam show enthusiasts. The first is the frizzle type, whereby each feather should curl towards the bird's head, and should be as tight and even as possible, whilst the second is similar to the stranded feathering of the Silkie breed (see page 89). The comb of all types should be single and evenly serrated with four or five points. There are at least 17 colours standardized and recognized in the US and almost as many in the UK, although the black-tailed white is the most often seen.

Breeding young birds is not always easy because mating and subsequent fertility can be difficult and the breed can also occasionally suffer from a genetic condition that arrests the development of the chick in the egg. Despite these potential difficulties, these pretty and very ornamental bantams are well worth considering by anyone looking for something a little unusual.

Jersey Giant

M

Another bird that was bred exclusively for the table is the Jersey Giant, which has its origins in North America, where the crossing of Brahmas and Croad Langshans (see pages 36 and 41) was instrumental in its development.

When the bird was first exported to Europe in the 1920s, only the black variety was seen, but today it is also possible to find white and blue-laced birds.

Bigger Jersey Giants are usually produced when stock is bred earlier in the season than is normal, giving the chicks the best of the summer months in which to grow. In days gone by, it was not unknown for the breed to live up to its name by producing caponized cock birds weighing as much as 9kg (20lb), although modern strains are nowhere near as heavy. At some stage, Australorps (see page 29) were also introduced into the bloodline, the result being that, size apart, the two look remarkably similar. Like most heavy breeds, the Jersey Giant makes a fine pet and it is also a good layer of brown eggs. However, it needs a fair amount of space in which to live, and the hens are not best suited to incubating and brooding their own chicks: although they make reliable mothers, because of their size they are prone to breaking their own eggs and accidentally squashing very young chicks.

In America, the breed is sometimes known as the Jersey Black Giant – not because of any colouring differences, but simply as an acknowledgement to the Black brothers who developed the breed in the 1870s.

La Flèche

fact file

Breed Type	Soft-feather. Heavy. Rare
Bantam	Yes
Purpose	Dual
Egg Laying	✔✔✔
Ease of Keeping	✔✔✔
Sociability	✔✔

> The most peculiar aspect of these birds is their strange comb, which is best described as two small, round horns growing upright and parallel to one another.

This old breed takes its name from the town of La Flèche in France. The most peculiar aspect of these birds is their strange comb, which is best described as two small, round horns growing upright and parallel to one another, and which has given rise to the breed's alternative names of 'Devil's Head' and 'Satan's Fowl'.

Although the French standard still requires that birds have a crest in contrast to other French poultry breeds such as the Houdan and Crève-Coeur (see pages 50 and 40), the British version of La Flèche has little or nothing in the way of crest, muffs or tassels. These features were bred out by English fanciers, using birds from some Mediterranean breeds in an effort to create a smooth head and white ear lobes: the resultant crosses unfortunately had the effect of reducing size and therefore usefulness as a dual-purpose bird. The La Flèche has never acquired a hugely strong following in the US, despite the fact that it was included in the first Standard in 1874.

The breed's original colour was black, but white, cuckoo and blue-laced varieties are now also accepted by the standards of most countries. As it was first produced as a table bird, La Flèche is a heavy breed. Despite this, the birds fly well and, if given the chance, often prefer to roost in any nearby trees rather than an allocated house – substantial covered runs may be necessary.

La Flèches are good layers of large white eggs, but in France they are still kept primarily as meat producers.

Lakenvelder

Originating in the Dutch/German border region (where its name is spelt Lakenfelder), this is a light breed with a single comb. The plumage patterning is a strong combination of black and white; in some books this is described as 'piebald' or 'belted' owing to the bird's white body and, in good specimens, solid black head and tail. The wing feathers should also have some black markings.

The Lakenvelder is not easy to breed to show standards, as there is a tendency for white feathers to appear in the neck – something that would be penalized by a judge.

Lakenvelders prefer to live free-range and can become quite nervy if they are penned in confined spaces. They are good layers of white eggs and only rarely tend towards broodiness.

Despite its nerviness when kept confined, the Lakenvelder is a hardy bird and, where space allows, will do well on free-range. Many keepers claim that their stock much prefers to forage for whatever is naturally available rather than to accept the easy hand-outs of regular feeding, but this is not a reason to assume that Lakenvelders can exist on natural food alone.

Due to the breed's scarcity in the US, it is very difficult to find prime examples and consequently virtually impossible to obtain quality breeding stock. The International Poultry Breeders Association (IPBA), based in Patterson, USA, offers an online database for Lakenvelder breeders throughout the world to create and maintain breeding records and ensure the future well-being of many breeds of poultry. Working with established and recognized poultry associations, the IPBA website was launched in August 2008.

The male Lakenvelder has noticeably more black feathering than the female, but both make striking show birds.

fact file

Breed Type	Soft-feather. Light. Rare
Bantam	Yes
Purpose	Layer
Egg Laying	✔✔✔
Ease of Keeping	✔✔✔
Sociability	✔✔✔

Leghorn

Leghorns have their origins in Italy (Leghorn is the German name for the Italian city of Livorno), but the development of the breed's various forms took place in a number of countries. Large exhibition types are bigger than those bred purely for utilitarian purposes, but even within the exhibition category there are extremes.

All possess a large single comb, but to conform to standards the comb of the hen should flop over in British strains, while that of American and Dutch-German types should be slightly smaller and only partially floppy. Rose-comb types are also allowed, despite the fact that they are rare. British Leghorns have small tails when compared with the Dutch-German large, full-feathered tail and the American spread tail.

The White Leghorn was imported to the UK via the US in the 1870s and the Brown Leghorn followed a few years later. Since then, many other colour variations have been developed. In the UK, black, white, brown, buff, blue and exchequer (a random mix of black and white) are acceptable; in the US, black, white, red, Columbian, partridge, brown, silver partridge and black-tailed red are now regularly seen; while in continental Europe, laced blue, red, yellow, bloodwing silver and cuckoo partridge types are included in the lists of standardized colours. Leghorns lay a plentiful supply of white eggs.

Interestingly, the bantam versions of the Leghorn are considered to be more docile than the large fowl and may well be worthy of consideration where space is limited or if they are being bought as family pets rather than purely as egg producers.

fact file

Breed Type	Soft-feather. Light
Bantam	Yes
Purpose	Layer
Egg Laying	✔✔✔✔✔
Ease of Keeping	✔✔✔
Sociability	✔✔✔

Malay

Another Asian hard-feathered breed, the Malay is also quite rare. Good specimens are tall with a long neck and legs, which are accentuated by the sparse plumage. The comb is compact and walnut-shaped, and the wattles are very small. The breed is described in some standards as having a 'cruel and morose expression', which is an appropriate definition.

This large fowl was first seen in the UK in the early 1800s, and for some reason it became particularly popular among breeders in Devon and Cornwall (possibly because of the need to introduce new blood into the Indian Game, see page 51). The size, shape and character of the Malay are more important than its colour. The shape of the bird is often referred to as 'triple-arched', whereby the first arch is formed by the neck, the second by the high-profile wings and the third by the tail, which is carried folded below the horizontal. The white and spangled colour varieties seem to be the most popular in the UK, but black, pile, red porcelain and duckwing are also found in other countries. The birds produce cream-coloured eggs but are not particularly good layers.

F

The size, shape and character of the Malay are more important than its colour. The shape of the bird is often referred to as 'triple-arched'.

fact file

Breed Type	Hard-feather. Heavy. Rare
Bantam	Yes
Purpose	Exhibition
Egg Laying	✔
Ease of Keeping	✔ ✔ ✔
Sociability	✔ ✔ ✔

Maran

Two of the most remarkable qualities of Marans are the dark, chocolate colour of their eggs and the fact that, although they are a heavy breed, they lay extremely well. Layers of dark eggs are, however, quite often bred commercially and so will suffer from the wrong leg colour, badly shaped combs and poor feather markings that exclude them from competitive showing and breeding. The breed originated in France, where nowadays birds often have some feathering on the legs.

Often known as Cuckoo Marans because of their colour, birds of this breed are actually found in several varieties, including brassy black, black, white, golden, silver cuckoo and Columbian. French fanciers also breed a wheaten-coloured bird. The comb is single and medium-sized. Marans are friendly towards people and also to members of their own species, so are suitable both as pets and for breeding purposes. The bantam version is ideal for those with only a small area at their disposal, but many bantam fanciers say that some strains fail to meet the depth of egg colour found in the large fowl. According to geneticists, this is because, when compared to the large bird, the bantam's egg has a bigger surface area, but the pigment gland that affects the egg's colouring is smaller and so, put simply, there is not enough dark pigmentation to entirely coat the shell.

The male French Copper Black and female Cuckoo, shown above, are just two of the many varieties of the Maran breed.

fact file

Breed Type	Soft-feather. Heavy
Bantam	Yes
Purpose	Dual
Egg Laying	✔✔✔✔
Ease of Keeping	✔✔✔✔✔
Sociability	✔✔✔✔

Marsh Daisy

The Marsh Daisy was developed around the turn of the 20th century in Lancashire, England, and became a proper, defined breed in 1913. It is believed to have Leghorn, Old English Game, Hamburg, Sicilian Buttercup and Malay in its make-up – resulting in a bird that is somewhat 'gamey' in character and an excellent forager in a free-range situation.

Surprisingly, despite being given sparse attention by its farming owners, the breed was found to lay very well throughout the winter, as well as being capable of producing a bird for the table which possessed an extraordinary amount of good quality breast meat.

Although the breed is recognized in five distinct colours of black, buff, brown, white and wheaten, the last is the most likely to be seen – in fact, such is the rarity of the breed as a whole, it is unlikely that you will come across any colour other than wheaten or brown. Marsh Daisies have distinctive large rose combs with a leader that follows the head, although not as closely as is some other breeds such as the Wyandotte (see page 102). The ear lobes are sometimes seen as being a combination of red and white, although the red pigmentation is generally dominant. Slow to mature and with a tendency towards flightiness, the breed's general hardiness, reasonable egg-laying ability and the knowledge that one is helping to ensure the future well-being of a very rare breed of poultry nevertheless make it worthy of consideration – provided, of course, that you can find a reliable breeder from whom to procure stock.

> The knowledge that one is helping to ensure the future well-being of a very rare breed of poultry makes it worthy of consideration.

fact file

Breed Type	Soft-feather. Light Rare
Bantam	No
Purpose	Dual
Egg Laying	✔ ✔ ✔
Ease of Keeping	✔ ✔ ✔ ✔
Sociability	✔ ✔ ✔ ✔

M

Minorca

The Minorca is similar in appearance to the Leghorn (see page 57) and, like that breed, is a soft-feathered, light bird from the Mediterranean (although the bantam version is a result of breeding in the UK and Germany). Colours are generally limited to black or white, but blues are standardized in most countries and are sometimes seen on the show bench.

The cock usually carries a large, erect, single comb; that of the hen should fold over. In some countries a rose comb is accepted, although in the UK the rose-comb bantam is not permitted as an exhibition breed. All birds should have large, white, oval-shaped ear lobes, but these sometimes suffer from spots and scabs. As the birds reach old age, the white colouring often spreads to the eyelids and face, making it impossible for them to be used for exhibition. The hens are excellent layers of large white eggs and, as the Minorca is also a pretty breed that needs little special attention, it has much to recommend it.

Although both the large fowl and the bantam are good layers and relatively easy to keep, the bantams in particular can be flighty and may need confining in a covered run. Also, in cold areas, it may be necessary to protect the comb and wattles with petroleum jelly to prevent frost damage.

> The cock usually carries a large, erect, single comb. All birds should have large, white, oval-shaped ear lobes.

fact file

Breed Type	Soft-feather. Light
Bantam	Yes
Purpose	Layer
Egg Laying	✔ ✔ ✔ ✔
Ease of Keeping	✔ ✔ ✔ ✔
Sociability	✔ ✔ ✔ ✔

M

Modern Game

This hard-feathered breed was developed in England as an exhibition bird some time between 1850 and 1900. It was derived from a cross between the Old English Game and the Malay (see pages 68 and 58) by a generation of judges and breeders who were too young to have had any contact with gamecocks (cock fighting was banned in the UK in 1849), who imagined – erroneously – that a bird with more reach would have had the advantage in a fight.

The breeders therefore selected for a longer necked bird, thinking this would look more 'authentic'. With the increasing popularity of showing, judges started looking for taller birds with a shorter hackle and smaller tails. The breed that was developed was originally known as the Exhibition Modern Game, but the word 'exhibition' was later dropped.

The comb of the Modern Game is single and small, and the breed is found in several colour varieties, including birchen, golden birchen, partridge, white bloodwing, silver bloodwing, black, blue and white. Its most notable features are its long legs, elongated neck and over-all carriage, which are all important if you intend to try your luck on the show bench. Modern Game birds are not brilliant layers, but because they are small and elegant and soon become very tame, they make excellent pets for the suburban poultry keeper.

fact file

Breed Type	Hard-feather. Light
Bantam	Yes
Purpose	Exhibition
Egg Laying	✔
Ease of Keeping	✔ ✔ ✔ ✔
Sociability	✔ ✔ ✔

Mid-19th century breeders thought that by producing a bird with a long neck they were re-creating the 'fighting cocks' of earlier generations.

Naine du Tournaisis

It will thrive equally well as a free-range bird or confined to a moveable house and run within the garden.

fact file

Breed Type	Soft-feather. True bantam. Rare
Bantam	Bantam only
Purpose	Exhibition/ornamental
Egg Laying	✔ ✔
Ease of Keeping	✔ ✔ ✔ ✔
Sociability	✔ ✔ ✔

It is said that this Belgian bantam was frequently kept onboard freight ships on the River Scheldt and, as a result, was given the colloquial name of either 'the Captain's Little Chicken' or 'Skipper's Little Chicken'. Originating on the French-Belgian border, it was at one time also known as the Mille Fleurs du Tournaisis and even today it is found only in the mille fleur colouring. It is, however, important to note that the colour markings can be irregular and are likely to differ from individual to individual; in addition, some pencilling may appear on the plumage of the female.

The Naine du Tournaisis ('naine' being the feminine form of the French word for 'dwarf') is clean-feathered and very attractive, being similar in looks to certain types of Old English Game bantams (see page 68), although the wings are slightly more slanted in appearance. It has a very alert head carriage, topped off by a moderately sized single comb.

Although Naines are most commonly seen in Belgium, Holland and northern France, it is possible to source stock in both the UK and the US, and so the breed might be worth considering as an ornamental pet. It will thrive equally well as a free-range bird or confined to a moveable house and run within the garden. In common with other exhibition or ornamental breeds of bantam, it is not a heavy or consistent layer, but what eggs it does lay are a good size in relation to the height and weight of the hen.

Nankin

Sometimes called Nankeens, this breed is thought to originate from the Nankin or Nanking region of China, although counterclaims maintain that it came from the East Indies and was named by traders because of its similarity in colour to a light reddish-yellow cloth that they used to export. Nankins are nowadays considered true bantams, but records show that both large fowl and bantams were originally imported to Britain in the 1700s.

The breed's overall appearance is alert, active and well balanced, the body being medium in length, moderately deep and carried well forward. Either rose or single combs are correct according to the standards of all countries. Sir John Sebright is believed to have used rose-comb Nankins in the development of his namesake breed (see page 85) and the two breeds are indeed comparable in size and stature. The Nankin's leg colouring is described in most standards as 'slate-blue', and the plumage as 'orangey-red to golden buff, with a black tail'. These little birds have a small but enthusiastic following of fanciers, all of whom praise their tameness, incubating ability and mothering skills.

fact file

Breed Type	Soft-feather. True bantam. Rare
Bantam	Bantam only (but see text)
Purpose	Ornamental
Egg Laying	✔
Ease of Keeping	✔ ✔ ✔
Sociability	✔ ✔ ✔

New Hampshire Red

Always a favourite in its native US, but also rapidly gaining popularity in the UK, the New Hampshire Red is thought to have been bred from the Rhode Island Red (see page 80) without the introduction of any other blood. They are, however, lighter in colour than Rhode Island Reds, though they too are classed as heavy and soft-feathered. The New Hampshire is probably best accepted as being dual-purpose, although it was originally bred as a layer.

Although the large-fowl type originated in the US, the New Hampshire bantam was developed in the Netherlands and should, in theory, be a true miniature of the large version in appearance. As well as the most commonly seen colour, chestnut-red, blue-marked and white examples are sometimes seen, although the standards adopted by some countries preclude the showing of any but the original chestnut-red colouring.

New Hampshire Reds are popular competition birds, especially in northern Europe, and are an excellent breed for the novice as a single colour (in this case chestnut-red) is easier to breed to exhibition level. They also have a placid nature and make perfect family pets, and as they are good layers they will keep a family in brown eggs for a good proportion of the year.

Another perfect breed for the beginner: handsome, productive and an easy-going family pet.

fact file

Breed Type	Soft-feather. Heavy. Rare
Bantam	Yes
Purpose	Dual
Egg Laying	✔✔✔✔
Ease of Keeping	✔✔✔✔✔
Sociability	✔✔✔✔✔

Norfolk Grey

The Norfolk Grey was created by Fred Myhill of Norwich, and was first shown at the 1920 Dairy Show under its original breed name of 'Black Maria'. Despite an enthusiastic reception, the fact that it was (and still is) a reasonable layer of light brown eggs and also capable of attaining a good carcass size for meat production, it never really became popular as a dual-purpose bird.

Its fortunes were such that in the 1970s it is believed that true breeding stock directly descended from those bred by Fred Myhill was reduced to a single quartet and, as is often the case in such situations, it is due only to the concentrated efforts of a handful of men that the breed is still in existence in the UK today. Still classified as 'rare', the Norfolk Grey has nowadays a dedicated following in both Britain and the US.

Although it is described as a heavy breed, the Norfolk Grey is not actually that large and weighs quite a bit less than others under the same classification. It has a single comb and is attractively marked: the cock bird has a silver hackle striped with black (as are the neck, back, saddle and wing feathers), while the rest of the plumage is black. The hen's hackle is similar to that of the cockerel, but otherwise she has black feathering all over, apart from the throat which is silver-laced.

It is due only to the concentrated efforts of a handful of men that the breed is still in existence in the UK today.

fact file

Breed Type	Soft-feather. Heavy. Rare
Bantam	Yes
Purpose	Dual
Egg Laying	✔ ✔ ✔
Ease of Keeping	✔ ✔ ✔ ✔ ✔
Sociability	✔ ✔ ✔ ✔ ✔

North Holland Blue

In Great Britain and America, the legs of the North Holland Blue should be feathered, but in their home country of the Netherlands the breed standards stipulate no feathers on the legs. The bird must always be blue-grey cuckoo in colour, making it very similar in appearance to the most commonly seen Maran colouring (see page 59). This detail can perhaps be best explained by the fact that both Marans and North Holland Blues are connected by a common ancestor, Malines or Mechelen Fowl from Belgium, and also share some Plymouth Rock blood (see page 76).

The North Holland Blue cockerel, like the Maran, is generally of a lighter colour than the hen, and it has a medium-sized red comb and red ear lobes. Introduced into the UK in the 1930s by Les Miles, it was considered a dual-purpose bird. Nowadays, although rare, it has a dedicated following of fanciers who appreciate its docile nature and ease of keeping. Excellent as free-range birds, North Holland Blues are known to become overweight if kept confined, a legacy of the days when they were as much prized for their table bird characteristics as for their egg-laying capabilities. In some countries, the breed is more correctly known as the 'North Holland Fowl'.

fact file

Breed Type	Soft-feather. Heavy. Rare
Bantam	Yes
Purpose	Dual
Egg Laying	✔✔✔✔
Ease of Keeping	✔✔✔✔✔
Sociability	✔✔✔✔✔

The bird must always be blue-grey cuckoo in colour, making it very similar in appearance to the most commonly seen Maran colouring.

Old English Game

The overall appearance, no matter in which country the birds are found, should be of a well-rounded, muscular bird, which is heart-shaped when viewed from above.

fact file

Breed Type	Hard-feather. Light
Bantam	Yes
Purpose	Exhibition/ornamental
Egg Laying	✔✔
Ease of Keeping	✔✔✔
Sociability	✔✔

Like the Modern Game (see page 62), the Old English has its origins in Britain and was tremendously popular as a fighting bird. Over the years some 30 colours have been known in this large breed, and in several countries any non-defined colour variations can be shown, as the judges are more interested in body shape and muscle tone.

The overall appearance, no matter in which country the birds are found, should be of a well-rounded, muscular bird, which is heart-shaped when viewed from above. Rumpless varieties are also seen. Traditionally, the small, single comb of the male birds was dubbed, or trimmed, to prevent injuries when fighting, but this practice is now illegal in many countries.

There is a very notable distinction between the feathering of Old English Game being bred in the UK and those seen in the US and Europe. In addition, there are colour complications and differences between the Carlisle and Oxford variations of the breed, perhaps the most obvious of these being the angle, or otherwise, of the back – the Carlisle's being more or less horizontal, whilst the Oxford's is at an angle of about 45 degrees. Up until the period between the two World Wars, all types were the responsibility of the Old English Game Club, but in the 1930s, the Carlisle and the Oxford Old English Game Clubs were formed as separate entities; the intention being that the Oxford club would look after the type of bird bred initially for fighting and the Carlisle concern itself with the 'exhibition' type.

Perhaps the main reasons for the popularity of the Old English Game are the showing and exhibition opportunities the breed offers – almost every club or poultry organization, no matter how small, has separate classes for all of the bird types and many of the colour variations. In Australia, however, the organizers of some shows separate birds into weight divisions rather than by any colour differences. The Old English Game Club of America looks after the well-being of the breed in general but the preference for showing bantams over large fowl has meant that there is also an Old English Bantam Club of America which exists to promote the bantam breeds and organizes almost 200 shows every year throughout the US.

Although Old English Game birds are aggressive towards each other, their boldness is a positive attribute when they are kept as pets, because they soon become tame. To help them stay fit and active, they are best kept free-range. Old English are probably the best layers of all the hard-feather breeds and produce a cream-coloured egg.

Old English Pheasant Fowl

Once known as the Yorkshire Pheasant Fowl, the Old English Pheasant Fowl was officially named in 1914, when a specialist breed club was formed. Photographs of birds from the early 1900s show very little change in the modern birds.

The breed today retains much of its original utility merits as a good layer of white or slightly tinted eggs (it was used in laying trials during the 1920s and 1930s), whilst producing a plump little table fowl. Like the Marsh Daisy (see page 60), it is a hardy breed, well suited to free-range and, provided that it can be protected from predators, needs little in the way of complicated housing. It is probably not a breed for those keeping poultry in a confined space.

The male bird has a moderately sized red rose comb, almond-shaped white ear lobes, rich bay and mahogany colouring with striped top and laced breast, slate-blue legs and feet, and a graceful carriage.

The hen is the same colouring, with crescent-shaped spangle markings (a quarter-moon-shaped spot of colour) at the end of each feather. The hens lay an average of 160 medium-sized white eggs each season but the pullets are known to be slow in maturing and do not normally come into lay until they are around seven months old. Unusually for a light breed, they will go broody periodically and make good mothers.

F

fact file

Breed Type	Soft-feather. Light. Rare
Bantam	No
Purpose	Dual
Egg Laying	✔✔✔
Ease of Keeping	✔✔✔✔
Sociability	✔✔✔

The breed today retains much of its original utility merits as a good layer of white or slightly tinted eggs.

F

Orloff

The Orloff is unusual in that it is one of the few poultry breeds to have originated in Russia, although the bantamized form was created in Germany in the early part of the 20th century. In fact, the large breed is now probably less popular amongst fanciers than the dwarf or miniature version.

American show standards recognize three colour varieties, but in the UK four are accepted, the best known of which is the mahogany (classified elsewhere as red porcelain). The Orloff has a walnut comb and a short, thick beak, and it stands tall on yellow legs. Both sexes have a full beard and bushy eyebrows. The birds are soft-feathered and come under the classification of a heavy breed – the large fowl was originally dual-purpose.

Orloff bantams would make interesting children's pets as they are very easy to hand-tame and do equally as well whether they are allowed free-range or kept confined – in which case, not being fliers, they do not need as high a fence as some other breeds. Although they are friendly towards their owners, the same cannot unfortunately be said of their attitude towards each other: when you buy birds to make up an unrelated group, there may be a little initial unpleasant-ness whilst a new 'pecking order' is established.

Perhaps as a result of their Russian origins, Orloffs are known to be hardy and are unfazed by inclement weather conditions. The hens lay a good number of cream-coloured eggs and, when broody, make reliable and caring mothers.

fact file

Breed Type	Soft-feather. Heavy. Rare
Bantam	Yes
Purpose	Dual/exhibition
Egg Laying	✔✔
Ease of Keeping	✔✔✔✔
Sociability	✔✔✔

Orpington

The Orpington makes a good beginner's bird and is popular throughout the world. When the black variety was first introduced at the London Dairy Show in October 1886, it produced quite a stir. Within five years it had been joined by white and buff varieties, and such was the demand that other breeders were crossing Croad Langshan and Cochin breeds (see page 41 and 39) and passing them off as Black Orpingtons.

Today, the four self-colours (buff, blue, black and white) are equally attractive and are almost the only variations ever seen in the UK. Elsewhere, birchen, buff-black laced, barred (or cuckoo), Columbian and partridge types are also quite common.

The Orpington lays a good number of light brown eggs. It has a small single comb, although the black variety sometimes has a rose comb. Its friendly, docile nature makes either the large fowl or the bantam an excellent choice of breed for the first-time chicken keeper.

Perfect though the large fowl Orpington is as a beginner's bird, the bantam is in a league of its own when it comes to providing the absolute children's pet. Cuddly and attractive, as well as being very easy to tame, there is the added bonus that although the cockerels will crow occasionally, they do not feel compelled to do so all day. Also, their crowing is nowhere near as loud as that of some other breeds, and so it may be possible to keep a breeding pen of birds in built-up areas without incurring the wrath of any close neighbours. The birds' conformation means that they have little if any inclination towards flying or rushing around and so they can be contained in a wire netting surround of no more than about 1m (3ft) (provided, of course, that they are not placed in any danger from predators or wandering dogs). Also of benefit when considering a children's pet is the fact that they will very easily go broody and make excellent and reliable mothers; so it should be possible to collect a clutch of eggs and place the broody hen in a suitable, safe, and secure place, giving young children the opportunity to see the whole incubation, hatching and rearing sequence from beginning to end.

The fact that the breed is very amenable and docile does, however, have one small disadvantage in that they may be subject to bullying when kept in a flock of mixed breeds.

> Its friendly, docile nature makes either the large fowl or the bantam an excellent choice for the first-time chicken keeper.

fact file

Breed Type	Soft-feather. Heavy
Bantam	Yes
Purpose	Dual/exhibition
Egg Laying	✔✔✔
Ease of Keeping	✔✔✔✔✔
Sociability	✔✔✔✔✔

F

Owl Beard

The most obvious difference between the Owl Beard (sometimes written as Owlbeard) and the Brabanter (see page 35) with which it is sometimes confused is that the Owl Beard has one continuous full round beard, while the Brabanter has a three-clump beard. Perhaps better known in Europe than it is in the US, the Owl Beard is currently enjoying popularity in the UK as well as something of a revival in Holland and Germany. Originating in Holland, the breed has been known there for several centuries and is believed by some to have been instrumental in the development of the Poland breed (see page 78).

The colour of the beak and legs in an individual bird depends on its plumage, which is many and varied. There are self-coloureds such as black and white, but many more appear as spangled, pencilled and even laced varieties. In addition, there is what might be best described as a 'sub-type' known as the Moorhead, which has its own colour variations of white, blue, golden and buff – all with a black head. The productivity of an Owl Beard apparently depends very much on its coloration, black or white types generally being considered to lay the largest number of eggs.

Owl Beards do better as free-range birds than in confinement, a factor that might influence their suitability for certain situations. In all other respects this is a hardy breed that is relatively easy to keep.

fact file

Breed Type	Soft-feather. Light. Rare
Bantam	No
Purpose	Dual
Egg Laying	✔ ✔ ✔
Ease of Keeping	✔ ✔ ✔ ✔ ✔
Sociability	✔ ✔ ✔

Pekin

There is sometimes some confusion between the bantam type of Cochin (see page 39), which has a large poultry counterpart, and the Pekin, which is classified as a true bantam by the Poultry Club of Great Britain. Both were imported from the Far East in the mid-19th century and some Pekin enthusiasts believe that the first birds were stolen from the private collection of the Emperor of China some time towards the end of the Opium Wars (around 1860). Others believe that today's Pekin is a derivative of the Cochin.

Whatever the truth of its origin, the breed was not accepted in its own right until the Birmingham Poultry Show of 1969 and in America only the Cochin is known. British standards stipulate that the Pekin should be single combed, and 'circular in shape when viewed from above, (the) whole outline rounded (with) heavily feathered legs and feet'.

The Pekin is a very attractive breed that looks well in a small garden, in which it can probably be allowed more freedom than some other birds becasue its heavily feathered feet prevent it from scratching and doing much damage to the flower beds. Conversely, the fact that it is so heavily feathered makes it less suitable for heavy clay soil and low-lying areas which are prone to wetness.

fact file

Breed Type	Soft-feather. True bantam
Bantam	Bantam only
Purpose	Ornamental/exhibition
Egg Laying	✔✔
Ease of Keeping	✔✔✔
Sociability	✔✔✔✔✔

Plymouth Rock

Another point in the breed's favour is that the birds are also very amenable in each other's company.

fact file

Breed Type   Soft-feather. Heavy
Bantam   Yes
Purpose   Dual
Egg Laying   ✔✔✔
Ease of Keeping   ✔✔✔✔✔
Sociability   ✔✔✔✔✔

This is an American breed, but the modern-day British type is taller and different in shape from birds found in the US. In the UK, the barred (thought to have been a result of breeding with the Scots Grey, see page 84) and buff varieties are perhaps the most commonly seen, but other colours such as partridge, Columbian, buff Columbian and blue-laced are also accepted. In the US the original Plymouth Rock has become known as the Barred Rock, and the breed has subsequently been further divided by colour.

As a point of interest, the barring in the US variety is broader than that seen in the UK. Similarly, in the US the black bars in the hens are broader than the white, while bars of equal width are desired in the cocks. The black should be beetle-green/black in colour and it should not be possible to confuse the birds with the Black Rock, which was developed as a commercial breed and is often found with a tinge of brown in its plumage.

Any of the Plymouth Rock variations would make a good choice for the novice. They are good layers of cream-coloured eggs (described as 'yellowish' in the US) and their placid nature makes them ideal for a relatively confined area: although, like most breeds, they appreciate being given free-range, they are almost as happy when penned. In addition, the fact that they are too heavy to even attempt flying means that any fencing considered necessary need not be all that high. That said, should you choose to include a soft netting 'roof' to prevent wild birds getting in and taking food (and, although this is a very faint possibility, introducing bird flu), the sides of the pen might need to be higher, purely for your convenience.

Another point in the breed's favour is that the birds are very amenable in each other's company and, provided that they are given sufficient space, it is possible to run more than one cock bird amongst a large group of hens. The hens, in common with most other soft-feathered, heavy breeds, are excellent sitters when broody and even better mothers, making the Plymouth Rock one of the best options for anyone new to chicken keeping and who wishes to hatch off chicks without the need for artificial incubators and brooders.

Poland

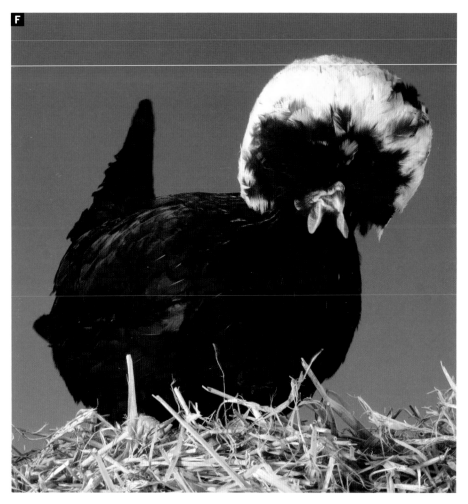

The Poland has a huge, ball-like feather crest and a small, sometimes completely absent, horned comb in front. Interestingly, its name is derived from the English word 'polled', rather than from the country of Poland.

As defined in the US, the UK, Germany, the Netherlands and several other countries, the breed is divided into two groups: the White Crested, with normal wattles; and the Paduaner, which is bearded. Nowadays, the colouring of the first group is restricted to black, blue or cuckoo-barred, while the latter group includes gold- and silver-laced, chamois, black, blue and white.

The colour variations and genetics of the breed make the production of future generations very interesting indeed. The White Crested, for example, has dominant genes and, if you were to mate it with one of the self-coloured varieties, most of the offspring would follow the White Crested in appearance. Even breeding self-colours is not without complications as, when mating a blue cock to a blue hen, there is the possibility that only half will be the desired colour – the remainder being a mixture of black and/or 'splash'. Mating a 'splash' to a black will, even more confusingly, most likely produce a hatch of black chicks. To achieve a better definition of barring in the cuckoo variety, the colour will be strengthened by the inclusion of a black. Likewise, to deepen the colour of a blue, it might be necessary to use a black from time to time.

> A little extra care should be taken to ensure that they are not unintentionally frightened by sudden movements.

The crest is as it is because the feathers that create it emerge from a protuberance at the top of the skull. Its exaggerated nature means that, in many individuals, the bird's vision is restricted and so a little extra care should be taken to ensure that they are not unintentionally frightened by sudden movements. On the plus side, the fact that they cannot see very well keeps them calmer than many other breeds.

To maintain the crests of Polands in good order, the birds should be kept under cover when it is raining and special drinkers should be used. If you have small pens of birds, nipple drinkers or aviary-type drinkers are best, but with larger flocks it is not a bad idea to use old-fashioned pottery pigeon drinkers or even dog bowls covered with a screen of wire netting.

While Polands do need special care, novices should not be put off by the potential extra work involved. The hens lay white eggs.

The Poland's remarkable crest is a large part of its attraction, although it is also found in a range of appealing colours.

fact file

Breed Type	Soft-feather. Light
Bantam	Yes
Purpose	Exhibition
Egg Laying	✔ ✔
Ease of Keeping	✔ ✔ ✔
Sociability	✔ ✔ ✔ ✔

Rhode Island Reds can lay more eggs than any other pure breed hen.

fact file

Breed Type	Soft-feather. Heavy
Bantam	Yes
Purpose	Dual
Egg Laying	✔ ✔ ✔
Ease of Keeping	✔ ✔ ✔ ✔ ✔
Sociability	✔ ✔ ✔ ✔ ✔

Rhode Island Red

These highly recommended birds are also excellent layers of light brown eggs.

Known simply as the Rhode Island in its native USA, this breed is generally given the suffix 'Red' in the UK and in other countries where a white variety is also found. In the red strains the comb can be either single (most commonly seen) or a rose, but in the white a rose comb is considered more correct. Confusingly, in the USA whites and reds are regarded as two different breeds. The red variety is much darker in colour than the New Hampshire Red (see page 65) but is otherwise very similar in stance and formation.

The Rhode Island Red originated as a result of crossing the Red Malay Game and other Asiatic stock, together with the Leghorn. Rhode Island Whites, on the other hand, are the result of crosses between Partridge Cochins, White Wyandottes and Rose-Comb White Leghorns. Some of the early breeders believed that the red variety was the better layer of the two colours, but there is no solid foundation to this belief. Somewhat bizarrely, Rhode Island Reds were exported to Europe as dual-purpose birds long before they were given standard recognition in their country of origin and, as a result, their genetic build-up can still be found in many of today's commercial hybrids. The bantam version was developed mainly in Germany and the UK, where they were created by mating large fowl Rhode Island Reds with other bantam breeds. Like the large fowl the bantam should conform to the 'brick-shape' appearance – that is, when viewing from the side, it should be possible to place its body into an imaginary square.

Rhode Island Reds are hardy and are happy to adapt to almost any healthy surroundings. These highly recommended birds are also excellent layers of light brown eggs.

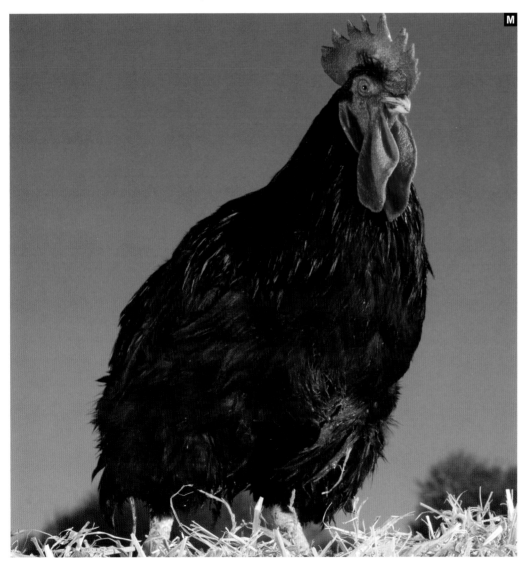

M *Their striking glossy plumage is a rich, chestnut red.*

Rosecomb

The Rosecomb is thought by many to be the oldest bantam breed in the UK and its existence was noted by writers in the latter part of the 1400s – in fact, it is known that King Richard III was a keen fancier of Rosecombs as a result of being given a pen of them by a Lincolnshire innkeeper. Although the breed is believed to have originated from Asia, the original British name of 'Black African' suggests that they arrived in the UK via something of a detour.

Rosecombs are very pretty, proud birds. They are full of character and, being small, make ideal pets in a limited space: they are perfectly happy confined to a moveable house and run. The most common colour is probably black, but whites can some-times be seen at shows and blues are becoming increasingly popular. In contrast to the limited number of colours seen in the UK, in America there are at least 20 possible variations, including barred, brown-red, buff, creole, exchequer, lemon-blue, mottled, silver duckwing and wheaten.

Despite being very easy to keep on a day-to-day basis, Rosecombs can be difficult to breed because some strains are subject to male infertility. If you are seriously intent on breeding, therefore, it will pay you to source stock carefully: try to gain some assurance from the breeder that the birds have not been subject to inbreeding in recent generations – a practice known to encourage hereditary faults.

fact file

Breed Type	Soft-feather. True bantam
Bantam	Bantam only
Purpose	Ornamental/exhibition
Egg Laying	✔ ✔
Ease of Keeping	✔ ✔ ✔ ✔
Sociability	✔ ✔ ✔ ✔

Rosecombs are very pretty, proud birds. They are full of character and, being small, make ideal pets in a limited space.

M

Scots Dumpy

This breed, whose name is sometimes spelt Scots Dumpie, is very old and may even have been brought to Scotland before the Roman invasion of Britain in AD 43. In its cuckoo colouring, the hen could perhaps be mistaken for a Maran (see page 59), although the cock bird is much finer and has a different shape altogether.

The Scots Dumpy has noticeably short legs which gives it a distinctive waddling gait quite unlike that of other chickens. The reason for their short legs and heavy body is traditionally attributed to the fact that Scottish crofts and smallholdings were surrounded by wild countryside in which predators were rife, so the birds were bred this way to encourage them to stay close to home.

At one time the Dumpy came in a range of colours, but unfortunately, owing to a loss of genetic material, only black, white and cuckoo varieties are seen today. Had it not been for a dedicated band of breeders in the 1970s, the breed could have died out altogether. The birds have a single comb, and the tail of the cock is full and flowing. The hen is considered to be an excellent broody and mother, and lays cream-coloured eggs.

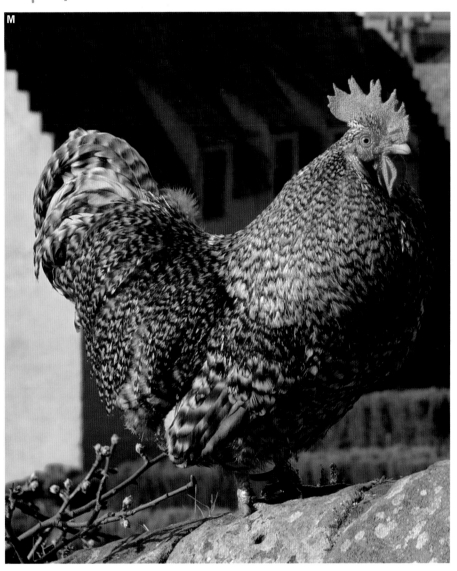

fact file

Breed Type	Soft-feather. Light
Bantam	Yes
Purpose	Dual
Egg Laying	✔ ✔ ✔
Ease of Keeping	✔ ✔ ✔ ✔
Sociability	✔ ✔ ✔ ✔ ✔

Scots Grey

fact file

Breed Type	Soft-feather. Light
Bantam	Yes
Purpose	Dual
Egg Laying	✔ ✔ ✔
Ease of Keeping	✔ ✔ ✔
Sociability	✔ ✔ ✔

At one time this breed was known as the Scotch Grey, but it is nowadays almost universally referred to by the name used here. It was popular in Scotland until around the 1930s, when it seems to have lost favour among chicken keepers in its own country. The breed standard recognizes only the cuckoo variety and insists that it should have a 'single comb' and 'white mottled legs', and be a 'fine, compact, smart bird with well-defined markings'. The Grey resembles the Scots Dumpy (see previous page) in appearance and its single comb is similarly accompanied by red ear lobes.

Although it is known as a hardy breed the Scots Grey is not, however, perhaps the best bird for a confined area. Its hardiness means that it requires space to free-range and, according to some breeders, is equally at home roosting in trees as in the chicken house. These factors are not that surprising when one considers the breed originates from Scottish farmyard birds well used to foraging and fending for themselves.

Sebrights are real characters and quite easy to keep.

Sebright

Sir John Sebright developed his namesake bantam at the beginning of the 19th century, making it one of the oldest British varieties of 'true' bantam. Somewhat unusually, the cock bird does not have the sickle and hackle feathers that so easily identify the male sex of many other poultry breeds and he is similar to the hen in having either gold or silver lacing throughout the feathering.

Occasionally, because of their genetic make-up individual males are infertile, and the intricate lacing required to comply with show standards may also be difficult to achieve. However, any negative points are far outweighed by the fact that Sebrights are real characters and, on a day-to-day basis, quite easy to keep – although because of their light, flighty nature, they will probably need some form of netted and fenced area. Even this is not a problem, as they will live quite happily in a combined house and run that can be moved around the garden from time to time, making them, like the Rosecomb (see page 82), an ideal choice for the small garden. The boiled egg for breakfast might, however, be a rarity, as Sebrights are generally expected to lay only between 60 and 80 creamy-white eggs each season.

fact file

Breed Type	Soft-feather. True bantam
Bantam	Bantam only
Purpose	Exhibition
Egg Laying	✔
Ease of Keeping	✔✔✔
Sociability	✔✔✔✔✔

Serama

Perhaps more correctly known as Malaysian Serama, the breed was developed in the Malaysian state of Kelantan and is supposedly the smallest breed of chicken in the world (as an indication of their height and weight, they range from between 15–25cm/6–10in and 225–335g/8–12oz). In the US, where competitions for the breed are becoming quite popular, they are graded into categories according to size rather than type or colour.

The Serama is variably known amongst its enthusiastic and increasing following as either the Arnold Schwarzenegger or the Dolly Parton of the ornamental bantam world due to its proud carriage and pugnacious appearance and heavily protruding breast!

For anyone who has limited space, yet wants to enjoy the experience of chicken keeping, Seramas might be the answer: their temperament, combined with the fact that a trio can be housed in nothing more than a decent-sized rabbit hutch, makes them ideal pets. Alternately, they would be perfect in the bottom of an aviary where they will thrive on the food dropped by other inhabitants. They do, however, generally require a warmer environment than the majority of the more usual true bantam choices and may benefit from being kept indoors or with access to a heat source (an ordinary light bulb suspended in their house will be sufficient) during the winter months. Seramas have a single comb and come in almost any colour variety. Interestingly, they do not breed true, and parents of one colour may produce chicks of several different variations.

M

fact file

Breed Type	Soft-feathered. True bantam
Bantam	Bantam only
Purpose	Ornamental
Egg Laying	✔
Ease of Keeping	✔ ✔ ✔
Sociability	✔ ✔ ✔ ✔

Shamo

The Shamo has been known for centuries in Japan, but its ancestors came from Siam (now Thailand). It was, until fairly recently, classified as rare by the UK Poultry Club. The fact that it is no longer on that list does not mean that its numbers are growing, merely that, along with several other hard-feathered breeds, it is now protected by the umbrella of a specialist club.

The breed is unusual in the UK in that even exhibition types have no colour standards. In Europe, however, a number of recognized colours have been standardized, including black, white, red porcelain, blue-laced, wheaten, and red- and silver-necked blacks. Three different comb types are allowed, all variations on a theme of small, compact, fleshy lumps. Shamo do not have wattles, but instead have dewlapped throats. The skin on the head and throat is smooth on young birds, but becomes thick and wrinkled after about two years of age.

The birds are easy to manage and, because the cocks are not known for crowing (and when they do, they don't make much noise), it is possible to have a breeding pen even where neighbours live in close proximity. They are not, unfortunately, gentle with one another, and if two cockerels were to get together there is a good chance that they would fight to the death. Like Indian Game males (see page 51), Shamo cock birds are virtually monogamous and should be bred in pairs or trios, kept separate in secure, portioned runs.

The Ko-Shamo is a true bantam and not a miniaturized version of the large fowl Shamo, although it is very similar in its overall appearance.

fact file

Breed Type	Hard-feather. Heavy
Bantam	Yes, but see text
Purpose	Exhibition
Egg Laying	✔✔
Ease of Keeping	✔✔✔✔
Sociability	✔✔✔

Sicilian Buttercup

Developed in Sicily, this breed was first seen in the US in 1835, but it was not until about 60 years later that any serious breeding stock was imported. The breed's arrival in the UK is reckoned to have been as a result of importation by a Mrs Colebeck of Yorkshire; a British breed standard originated in 1920.

The Sicilian Buttercup, in common with many other Mediterranean breeds, was used extensively in laying trials during the 1960s, for its ability to produce very high numbers of eggs when fed on the most basic of rations.

The name 'Buttercup' supposedly comes from the bird's very unusually shaped comb, described as 'a cup-shaped crown with a complete circle of medium-sized regular points' and resulting from a genetic variation of two side-by-side single combs that eventually merged at the front and rear. Another notable characteristic of the breed is the vast difference in colouring between the sexes of the gold-spangled variety: males are a deep orange-red with a black tail, the cape showing some black spangles, whilst hens are golden-buff, with parallel rows of black spangles. There is a silver-spangled variety, but this is seldom seen, and colours such as brown, white and gold-duckwing, recorded when the breed was first imported from Sicily, are nowadays unknown. Ear lobes are red in the UK standard, but white in the US. The leg colouring, like that of the Derbyshire Redcap (see page 42), is willow-green. The Sicilian Buttercup is sometimes erroneously referred to as the Sicilian Flower-bird.

The name 'Buttercup' comes from the bird's very unusually shaped comb.

(see page 42)

M

fact file

Breed Type	Soft-feather. Light. Rare
Bantam	Yes
Purpose	Laying/exhibition
Egg Laying	✔ ✔ ✔
Ease of Keeping	✔ ✔ ✔
Sociability	✔ ✔ ✔

Silkie

Silkies are a light breed that arrived in the UK from China sometime in the early 1800s. Since then, they have become extremely popular particularly as children's pets – there is no doubt that their unusual feathering fascinates youngsters. Colours include white, black, partridge, cuckoo, red-buff and lavender, although not all of these are recognized everywhere.

The comb is known as a mulberry or cushion, and is almost circular in shape but preferably broader rather than long. The male has a slightly spiky crest, while that of the female should be short and neat, resembling a powder puff. Some strains also have a beard. Both males and females have an extra toe at the back of each foot. Also typical of the breed is the blue-black skin colour.

Silkies are docile and easy to keep, but because of their feathering they should always have access to a dry run and warm housing. They lay a creamy-brown egg and can be expected to produce an average of 105 eggs before going broody. The breed is particularly useful for crossing with others to produce the ideal surrogate mother. Some chicken and bantam breeders keep a pen of Silkie crosses to use purely as sitting hens for clutches of eggs coming from their main-interest breeds which, for a variety of reasons, are not likely either to go broody themselves, or make poor mothers when they do. The reason a cross, rather than a pure-bred Silkie, is preferred is that the crosses have all the superb mothering tendencies of the Silkie but lack the fine, cotton-like feathering, so are less likely to strangle the brooding chicks.

fact file

Breed Type	Soft-feather. Light
Bantam	Yes
Purpose	Ornamental/exhibition
Egg Laying	✔
Ease of Keeping	✔ ✔ ✔
Sociability	✔ ✔ ✔ ✔ ✔

Spanish

Also known as the White-faced Black Spanish, this Mediterranean breed is, at first glance, similar to the Minorca and Leghorn (see pages 61and 57) although on closer inspection it may be noticed that it is slightly longer and more sloping along the back. It also has a single comb and large white wattles. Unlike the Minorca and Leghorn, the Spanish must be black if it is to comply with breed standards.

Developed over many years purely as laying birds, the females very rarely, if ever, go broody – so they are ideal for anyone who requires a breed solely to lay eggs. Unfortunately, like many Mediterranean types, the Spanish is quite noisy, flighty and excitable, making it difficult to keep either in a confined space or in areas where neighbours have to be considered. The bantam version is somewhat more placid and therefore might be a suitable choice in such situations.

Although the Spanish was developed as a laying bird, some fanciers have, over several generations, selected breeding birds with a view to accentuating the large white ear lobes and facial skin, with the result that this is now considered an ornamental breed in some countries. The white parts are of particular interest in that, although they are present at hatching, it is not until the bird reaches a year old or more that these quite striking features are fully formed.

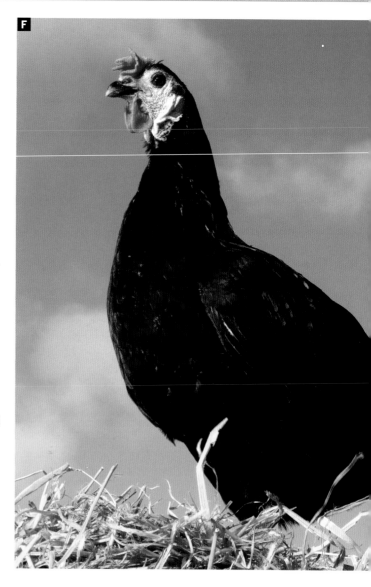

F

The Spanish is quite noisy, flighty and excitable, making it difficult to keep either in a confined space or in areas where neighbours have to be considered.

fact file

Breed Type	Soft-feather. Light. Rare
Bantam	Yes
Purpose	Layer
Egg Laying	✔ ✔ ✔ ✔
Ease of Keeping	✔ ✔ ✔
Sociability	✔ ✔ ✔

Styrian

The Duchy of Styria lay between southern Austria and northern Slovenia, and it was from here that Styrian poultry originated. Its characteristics suggest a common ancestry with some of the Mediterranean breeds, but although first mentioned as early as the 13th century, it merely existed until the second half of the 19th, when breeders began to realize its potential as a dual-purpose bird.

Their enthusiasm was, however, short-lived and, even though the Styrian was given a great deal of emphasis at various important symposiums which dealt with the future of many breeds of poultry, there was, by the end of World War I, very little interest in it. Fortunately, stock was collected in order that a preservation programme could be begun and this must have been moderately successful as, in 1930, a breed standard was defined. World War II prevented any further significant progress on improving numbers and standards, and it was not until 1987 that any serious efforts were made in those quarters.

The hens are quick to mature, lively and quite good layers of medium-sized, ivory-coloured eggs. An adult cock has a small, fine-shaped head, the comb being of medium size, behind which can often be found a tassel of thin tufts of feathers. Four varieties of colour are known: light brown, white, barred and partridge, the last of which is the most commonly seen.

fact file

Breed Type	Soft-feather. Heavy. Rare
Bantam	No
Purpose	Dual
Egg Laying	✔ ✔ ✔
Ease of Keeping	✔ ✔ ✔ ✔
Sociability	✔ ✔ ✔ ✔

Sultan

For anyone looking for something really unusual – and with a royal pedigree to boot – the Sultan is well worth considering. It was, as its name suggests, kept by sultans of the Turkish courts (although none of today's stock is likely to be descended from these birds) and its heavy crest, beard and muffs and heavily feathered legs would certainly make an interesting talking point as it busied itself around the garden.

Unfortunately, however, the Sultan is definitely ornamental rather than practical. The very features that make it so attractive render it unsuitable for all but the driest ground, though it could, of course, be kept in an enclosed house and run – a situation to which it is particularly well suited. A covered run would also prevent individual specimens from becoming 'brassy' as a result of being out in the sunshine or extreme weather – an undesirable trait that would prevent a bird from being shown until the next moult (brassiness can be a great problem to the showing enthusiast of any breed that includes white, lavender or other light-coloured birds).

In the US, black, blue or white coloration is recognized, but in the UK only white is acceptable. Another unusual feature of the breed is that, like the utilitarian Dorkings (see page 43), Sultans have five toes. The small comb, which is almost completely obliterated by the heavy crest in the cock bird, should be of the V-shaped horn variety.

The potential problems involved in the Sultan's keeping ought not to deter the would-be fancier, especially as the breed is known for its friendly temperament.

fact file

Breed Type	Soft-feather. Light. Rare
Bantam	Yes
Purpose	Ornamental
Egg Laying	✔
Ease of Keeping	✔ ✔ ✔
Sociability	✔ ✔ ✔ ✔ ✔

> If you are tempted to keep Sumatras, bear in mind you should confine them in large grassy runs.

fact file

Breed Type	Soft-feather. Light. Rare
Bantam	Yes
Purpose	Exhibition
Egg Laying	✔ ✔
Ease of Keeping	✔ ✔ ✔ ✔
Sociability	✔ ✔ ✔

Sumatra

These are slender, graceful, striking birds with a dark blue triple or pea comb – this colouring is referred to by some breeders as 'gypsy'. Red combs, faces and wattles are sometimes seen, but they are not desirable on the show bench. Generally, only two colours are seen – blue and black – but in Germany a white strain was developed by mating a Sumatra with a Yokohama (see page 106). Elsewhere, there is also a black-red variety.

As well as the unusual 'gypsy' colouring of the comb, face and wattles, the Sumatra stands apart from most other breeds by dint of the fact that the male bird grows two or more spurs at the back of each leg. The tail feathering is unusual in that, although very luxuriant, the sickle feathers are positioned in such a way that they keep the main body of feathers from trailing on the floor. However this breed would not be suited to wet, muddy conditions and their long tails

necessitate large houses with high perches.

If you are tempted to keep Sumatras, bear in mind you should confine them in large grassy runs or, better still, give them total free-range. Be aware, however, that they may prefer to roost in any surrounding trees rather than in their allocated home. Unusually for poultry that originates in Asia, the Sumatra lays white eggs.

The well-known Sussex, which developed from the Speckled Barn at around the time of the Roman invasion of England in AD 43, was originally a table bird but is now dual-purpose. Outbreeding with Cochin blood (see page 39) probably produced the first Light Sussex and as a result the colour of the eggs became tinted and the numbers produced greatly improved.

Some fanciers do not regard the Light Sussex as a true variety, and in a number of countries it is known by its colour, Columbian. The breed also recognizes six other types: speckled, buff, white, silver, brown and red. Cuckoo variety is allowed in some parts of the world.

The plumage variations of the Sussex can be difficult to achieve to show standard, but that should not put anyone off trying one of the less common colours. Note also that if the lighter shades are being shown they need to be kept out of sunlight to prevent them from turning brassy. The breed has a single comb and a good, solid body, making it full of character and giving it the look of a typical 'farmyard' bird. The hens are excellent layers despite being classified as a heavy breed. Along with the Rhode Island Red (see page 80), the Sussex, and particularly the Light Sussex, has been a favourite amongst both farmers and backyard poultry keepers in the UK for many years, particularly during World War II when it was kept extensively as part of the 'war effort', providing eggs and meat in households that would otherwise have lacked both commodities. Originally, however, the Light Sussex was primarily a table bird: so much so that a whole industry grew up around the areas of Tunbridge Wells and Eastbourne in order to supply the London meat markets on a daily basis. With subsequent breeding, much of the meat-production aspects of the bird's make-up have been lost in favour of a lighter, egg-laying breed and nowadays both hybrid-sized and much larger traditional types are to be seen. Those kept for exhibition seem to be more frame than bulk and so, should a table bird be required, it is important to acquire the right stock.

Like the Rhode Island Red, the Light Sussex has been widely used in the development of sex-linked hybrids for commercial use, but do not let this fact be confused with the oft-quoted belief amongst poultry fanciers that it is possible to sex chicks of the Speckled Sussex type by assuming that the cocks will have white wings and the pullets light brown wings, as there is no proven scientific basis for this supposition.

fact file

Breed Type	Soft-feather. Heavy
Bantam	Yes
Purpose	Dual
Egg Laying	✔ ✔ ✔ ✔
Ease of Keeping	✔ ✔ ✔ ✔ ✔
Sociability	✔ ✔ ✔ ✔ ✔

Transylvanian Naked Neck

At first glance, the Transylvanian Naked Neck looks like a bird that has spent its life with its head and neck through the bars of a battery cage: virtually all the neck and crop area is bare of feathering.

The scientific reason for this is that any reduction in feather mass improves heat dissipation through the naked area, improving tolerance to heat and leading to greater productivity under high ambient temperatures. These were important attributes in the extreme heat of Eastern countries, where the breed was developed to fulfil a need for commercially viable table birds.

When the breed was first introduced into Great Britain in the 1920s, the media apparently made much of the fact that the birds were a result of a mating between chickens and turkeys and for this reason they were called Churkeys. Elsewhere, they acquired the name Turken for much the same reason. In the US, the prefix 'Transylvanian' is not used and the birds are known simply as Naked Necks.

Apart from the lack of neck plumage, the breed is similar in shape and conformation to most other utility birds. It has a single, deeply serrated, medium-sized comb and the feathers of the large fowl can be blue, buff, cuckoo, red or white. Leg coloration is dependent on the plumage colouring.

By no stretch of the imagination can this breed be called attractive, but there is no doubt that its extremely unusual appearance has ensured an enthusiastic following in some quarters.

The 'naked' neck improved this breed's ability to tolerate the extreme heat of its native lands.

fact file

Breed Type	Soft-feather. Heavy. Rare
Bantam	Yes
Purpose	Table/dual
Egg Laying	✔✔✔
Ease of Keeping	✔✔✔✔
Sociability	✔✔✔✔

Tuzo

Much of what has been written about the Shamo (see page 87) applies to its cousin the Tuzo (sometimes known as the Nankin-Shamo just to confuse matters further).

There is some discrepancy as to the breed's true origins as, according to *Cockfighting all over the World*, a weighty tome written in 1928 by Carlos Finsterbusch, the Tuzo was once bred and owned by members of the Japanese royal family and their courtiers yet the Japan Poultry Society claim that no breed has ever been known as a Tuzo in their country. There is, however, a possibility that small specimens of the Shamo were known colloquially as 'Tuzi' – a name that stuck when American soldiers took examples of the breed back to the US after the cessation of their war service in Japan. The Tuzo's appearance in Europe came about as a result of American birds being imported to Holland in the mid-20th century, but it was not until the 1970s that it became known in the UK.

The importation of several very differing strains of the breed into Europe has meant that some examples follow the Shamo in appearance whilst others are more like the Aseel (see page 27) in general outline. Typical features include a triple (pea) comb, virtually non-existent red wattles, a very upright and aggressive-looking appearance, and very strong thighs and shanks. Plumage coloration is many and varied, including black-red, brown-red, creole, cuckoo, pile, wheaten, and both gold and silver duckwing.

> The Tuzo's appearance in Europe came about as a result of American birds being imported to Holland.

fact file

Breed Type	Hard-feather. True bantam. Rare
Bantam	Bantam only
Purpose	Ornamental
Egg Laying	✔✔
Ease of Keeping	✔✔✔
Sociability	✔✔✔

M

Twente

Twente, or Twense, is the Dutch name for a breed that is also known by its German name of Kraienköppe and was developed on the borders of the two countries. Although not well known in the UK, where it is classified as a rare breed, the Twente is popular in its countries of origin.

The birds are good-looking, with a proud, erect stance, not unlike that of the Old English Game (see page 68). The comb is walnut-shaped in the male but hardly exists in the female. Colours are partridge, silver partridge, blue partridge and silver-blue partridge, although in the UK the two varieties known are called gold and silver.

The birds are robust and hardy, and soon become tame towards humans, but unfortunately are sometimes aggressive towards each other. The hens are good layers. Both large fowl and bantam types lay cream-coloured eggs and the large fowl are known for laying right through winter, unlike some other breeds. The Twente bantam makes a good mother and will successfully brood and hatch a clutch of seven to nine eggs. Twentes are good fliers and so will require high fencing or, better still, a netted-over run. Keeping a breeding pen, which obviously necessitates the inclusion of a cockerel, is not a good idea in a built-up area, as they are noted for their inordinately loud crowing.

fact file

Breed Type	Soft-feather. Light. Rare
Bantam	Yes
Purpose	Layer
Egg Laying	✔✔✔
Ease of Keeping	✔✔✔✔
Sociability	✔✔

Both large fowl and bantam types lay cream-coloured eggs and the large fowl are known for laying right through winter, unlike some other breeds.

Left: The handsome Vorwerk owes its survival to the commitment of a handful of dedicated fanciers.

Vorwerk

This breed originated in Hamburg in 1900 and was the result of work by Oskar Vorwerk, who required a bird that would be not only dual-purpose, but also economical to keep and rear. Vorwerks were recognized as a pure breed in Germany in 1919, but much of the original stock was lost during World War II and the breed had to be reconstituted from what birds remained. Vorwerks were not imported into the UK until very late in the 20th century and then only as a result of an interest by a Mrs Wallis of Arundel, West Sussex.

The breed is only found with black-belted markings on a buff-coloured body and, apart from the base colouring, it is often described as being similar in plumage, shape and stance as the Lakenvelder (see page 56). Like the Lakenvelder, the Vorwerk has a single comb and its legs are slate-blue.

In America, the bantam version of the breed was developed by Wilmar Vorwerk (who, despite having the same surname, is no relation of the original breeder) by crossing and culling a mixture of bantam breeds until a bird that bred true was obtained. The bantam was recognized by the American Bantam Association standards in 1985. Bantams and large fowl alike are quite flighty and require reasonably high fencing to confine them, but in all other respects they are an easy breed to keep.

fact file

Breed Type:	Soft-feather. Light. Rare
Bantam:	Yes
Purpose:	Dual
Egg Laying:	✔ ✔ ✔
Ease of Keeping:	✔ ✔ ✔ ✔
Sociability:	✔ ✔ ✔ ✔

Welsummer

This Dutch breed takes its name from the village of Welsum and has blood from the Cochin, Wyandotte, Leghorn, Barnevelder and Rhode Island Red breeds in its make-up. In most countries the Welsummer is found in only one colour, which in the male is similar to that of a black-red Old English Game (see page 68) and in the hen to that of a Brown Leghorn (see page 57).

Unlike other black-red birds such as the Brown Leghorn and Partridge Wyandotte (see page 102), which have pure black breasts, the Welsummer has a breast that is a mixture of brown and black, similar to that of a pullet-breeder Partridge Wyandotte. The birds have a single comb.

Compared to other egg-laying breeds, Welsummers are relatively poor layers, but the dark brown colour of their eggs is comparable to that of the Maran (see page 59), the only difference being that the shell of the Maran is glossy whereas that of a Welsummer is matt. Interestingly, the darkest eggs often come from poor layers, while the better layers produce lighter coloured eggs. Therefore if the breeder selects hatching eggs that are deep chocolate in colour, he may unwittingly also be selecting for low production in the future. Although no shell colour affects the quality or otherwise of an egg's contents, the fact that the coloration of a Welsummer's egg is such an important feature of the breed should never be forgotten, and every effort must be made to maintain the dark chocolate pigmentation. Indeed, the Welsummer Club has a separate egg standard to be used when judging eggs of the breed. When selecting possible breeding stock purely as a result of the eggs a hen produces, it is as well to bear in mind the fact that, as an individual bird grows older, the eggs she lays may well become paler; therefore, if as a pullet she laid dark eggs that were typical of the egg standard, she should still be considered. In addition, one should always select breeding males from hens who are known to lay deep-coloured eggs.

The Welsummer is an ideal breed for the newcomer to chicken keeping as it is very friendly and will thrive equally well whether it is kept free-range or in a closed run. It is thrifty in its eating habits and is very hardy making it far less likely to succumb to many of the health problems sometimes associated with other poultry breeds. Bantam versions are more lively and active: when kept in an enclosed run, they should be kept occupied by means of daily titbits, or by the addition of a turf of grass on which they can peck and scratch.

fact file

Breed Type	Soft-feather. Light
Bantam	Yes
Purpose	Layer
Egg Laying	✔ ✔ ✔
Ease of Keeping	✔ ✔ ✔ ✔ ✔
Sociability	✔ ✔ ✔ ✔ ✔

Wyandotte

The soft-feathered, heavy Wyandotte breed originated in the US as the Silver Laced Wyandotte. The feather lacing came about by crossing a Silver Sebright bantam with either a Cochin or a Brahma and then introducing a Silver Spangled Hamburg. The white variety is a sport of the Silver Laced Wyandotte and was developed in the UK before being exported back to the US, where it was further improved to become an excellent layer of white or cream-coloured eggs.

The breed has at least 14 colour variations in the UK, and at least 22 in total around the world. It is not, however, just the variety of colours or the general appearance of the birds that makes them so popular, but rather their ease of keeping. Some of the colour variations do make for complicated breeding problems – the partridge variety, for example, is produced through the process of double mating (see page 17), for which separate pens of pullet- and cock-breeding birds are required. Like many other pale-coloured breeds, the White Wyandotte needs protection from the sun if you intend to exhibit it, but even so it is perhaps the best variety for the beginner.

Another of the Wyandotte's great attractions is its rounded, fluffy-looking shape: it is often described as being 'a bird of curves' – each of these curves being very clearly defined, especially in the head, tail and body sections, every one of which should fit within an imaginary circle. It has been said that the thickness and density of soft feathers found in the vent area makes mating difficult and that fertility is sometimes affected as a result; where or why this idea originated is not clear, and it is the opinion of the majority of experienced Wyandotte fanciers that this 'fact' should be ignored.

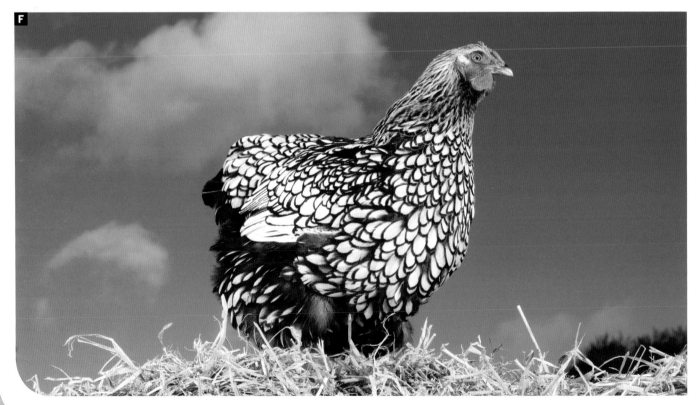

F

The dense feathering might, however, cause some concern in respect of keeping the vent area clean of faeces; although this is rarely a problem in a healthy bird, a close watch should be kept on any build-up in that area as, left unattended, it will very quickly become a breeding site for mites and parasites.

During its development and because of the many and varied breeds used before the breed became standardized, the Wyandotte has had several types of comb, but the modern bird should only ever be seen with a medium-sized rose comb which is close-fitting to the head and not so wide that it hangs over the eyes. The legs should be clean, thick-fleshed and yellow – although the leg colouring may appear lighter on older laying hens. The standard requirement for yellow legs can cause a problem to the potential breeder of some Wyandotte types – some strains of the black, for instance, can show a preponderance of dark scales rather than yellow. Despite any of these possible difficulties, the docile, hardy, friendly and attractive Wyandotte cannot be recommended highly enough as being the ideal beginner's breed.

fact file

Breed Type	Soft-feather. Heavy
Bantam	Yes
Purpose	Dual
Egg Laying	✔ ✔ ✔ ✔
Ease of Keeping	✔ ✔ ✔ ✔ ✔
Sociability	✔ ✔ ✔ ✔ ✔

The docile, hardy, friendly and attractive Wyandotte cannot be recommended highly enough as being the ideal beginner's breed.

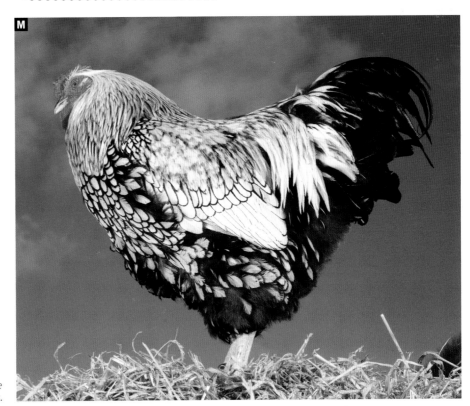

'A bird of curves': Wyandottes of both sexes are much softer in outline than most other breeds.

Yokohama

The long-tailed Yokohama cannot be mistaken for any other breed. In a mature bird the tail can easily reach 60cm (2ft) in length, and in Japan (where the breed originated) fanciers keep the birds in conditions that prevent them from moulting, to encourage the tail to grow as much as 1m (3ft) each year. Both single and walnut combs are accepted in the breed standards. The former is most often seen in black-red and duckwing colorations, while the latter is more common in red-saddled and white varieties.

To describe how the Yokohama arrived into Europe and beyond is to explain how the breed came to be called as it is. In Japan, there never was a breed known as Yokohama; the birds have simply taken their name from the port at which they were first exported by a French missionary called Girad. As well as France, birds found their way into Germany, where they were further developed by the addition of bloodlines from another long-tailed Japanese breed known as the Minohiki, the resultant offspring being eventually taken to the US.

Keeping Yokohama cock birds in prime condition can be difficult; to maintain perfect show specimens you do need taller houses and higher perches than normal. The laying capabilities of this light, ornamental breed can be quite varied. Depending on the strain, some hens will lay only 40–50 eggs in a season, while others can average somewhere around 100.

fact file

Breed Type	Soft-feather. Light. Rare
Bantam	Yes
Purpose	Ornamental/exhibition
Egg Laying	✔
Ease of Keeping	✔ ✔
Sociability	✔ ✔ ✔

Yurlov Crower

Like the Orloff (see page 71), the Yurlov Crower originated in Russia. As its name suggests, it was used in the crowing competitions which were extremely popular throughout Russia at the beginning of the 20th century and was probably developed from table birds kept by peasants and farmers.

According to records and website information, there were, even as late as 1985, only 200 specimens of Yurlov poultry registered throughout Russia and several of these were kept at government poultry farms, without which the breed would undoubtedly have been in an even more parlous state. Fortunately, the population in Russia and the Ukraine now numbers thousands rather than hundreds, but it is probably due to an interest in the breed by a few German poultry fanciers that the Yurlov Crower has come to Europe – and this only in the last 20 years. Colour variants are black, white, silver, golden, scarlet-red or black with light yellow, and the breed can nowadays be seen with either a single or a rose comb. Despite being a heavy bird, it is nevertheless a relatively good layer and must therefore be classed as dual-purpose.

Whether the male's crow differs from that of other breeds because it was bred for use in crowing competitions or whether its very different crow made it suitable for competitive purposes is not made clear in contemporary records, but it is a fact that the Yurlov cocks were entered in classes according to their tone.

fact file

Breed Type	Soft-feather. Light. Rare
Bantam	Yes
Purpose	Dual
Egg Laying	✔ ✔ ✔
Ease of Keeping	✔ ✔ ✔ ✔
Sociability	✔ ✔ ✔ ✔

M

HYBRIDS

A hybrid is the result of a cross between two or more breeds. Such a bird will not breed 'true' and reproduce chicks in its own likeness (for more of the same type you would have to go back to the original parent stock and breed from them again), but on the plus side hybrid hens tend to be more vigorous and productive than pure breeds.

Hybrid hens can lay anywhere up to 280 eggs per annum. and the vast majority of modern commercial birds are hybrid stock, though many of the types available have names more reminiscent of soft porn actresses (Amber Star) or Wild West movies (Gold Star Ranger) than anything connected to the poultry world! Hybrid laying poultry have, over the years, been produced in different ways, but have always been based on parental strains selected to complement one another genetically: a procedure known by some commercial producers as 'nicking'. It was the practice to test a number of different strains so that the best could be used for producing the commercially marketable end product.

Because producers provide many varieties on a theme, similar-looking hybrid layers may be offered for sale under a multitude of names. With feather colouring including gold, black, silver, white, speckled and blue, it would, in fact, be possible to 'mix-and-match' and colour co-ordinate your poultry flock with the flowers in your garden!

The majority of hybrids have been developed from the traditional poultry breeds: a cross between a Rhode Island Red and a Light Sussex is a common one, as is that of a Rhode Island and a Barred (Plymouth) Rock. These matings are also sex-linked and, depending on the parentage, it should be possible to identify the sex of the chicks at a day old. Mating a Rhode Island Red cockerel

Above: The colour of an egg yolk is directly associated with the hen's diet: a regular, natural supply of greenstuffs will almost always produce a deeper yellow yolk.

with a Light Sussex hen, for example, will produce females of a golden/buff colour and males of a silver/white. Adult plumage will also vary depending on the way a mating is made: 'Black Star' hybrids are created using a Rhode Island Red cockerel with a Barred (Plymouth) Rock hen, whereas, putting a Barred Rock male with a Rhode Island female will produce what is commercially known as the 'Speckled Star'.

Best known of all egg-laying hybrids in the UK is probably the Black Rock – predominantly black, but with chestnut colouring around the neck – which can be

Left: Some hybrid types may not be exciting to look at but they are all excellent layers.

bought from many agents throughout the country but is bred only by Peter Siddons at Muirfield Hatcheries in Scotland and is ideal for the more inhospitable parts of the country where other commercial breeds may not fare so well. In America, the nearest equivalent is the Bovans Nera, but the company Aviagen produces three other well-known hybrid strains: Arbor Acres, Ross and L.I.R.

Traditional hybrids bred for the table were created by the use of the White Rock and Cornish Game, the genes of which still feature prominently in today's meat-producing birds as they score over the pure breeds for their ratio of meat to bone, their speed of growth and their ultimate weight. Unfortunately, their quite phenomenal growth rate has its disadvantages in that they sometimes suffer from leg and heart problems, which can be fatal. Perhaps the best table bird for the hobbyist is the Sasso – a breed originating from France that combines hardiness and ease of rearing with a fine flavour; it will forage and roost quite happily on free-range. Much useful information can be found on the Sasso company website: www.sasso.fr

Left: The Black Rock is a particularly hardy type, a fact that makes it most suitable for keeping free-range.

RESCUE & EX-BATTERY BIRDS

Fifty years ago, apart from small-scale producers using a semi-free range system, almost all commercially produced eggs came from hens kept in the battery system. They were, according to the thinking of the time, 'ideal where space is limited... there is no denying that production from them equals, and very often exceeds, production from any other types of accommodation.'

Although battery hen types are given names such as Goldlines, ISA Brown's and Warrens, they are nevertheless all birds bred specifically to produce the largest number of eggs possible. Interestingly, brown egg layers are the preferred option for UK egg producers; whereas in the US it seems that the egg-buying public much prefers white-shelled eggs. There has always been the opportunity to buy ex-battery hens quite cheaply as soon as they come to the end of their first proper laying season; the majority of commercial egg producers replace them with pullets once the moult has begun. Unfortunately, until relatively recently, few people showed any interest in these birds, even though they would have had several good egg-laying seasons in front of them, with the result that perfectly good hens were killed off to supply chicken pies and pet foods. Nowadays, thanks mainly to the huge efforts made by the Battery Hen Welfare Trust, a charity set up by Jane Howorth in 2004, the ex-battery hens of the UK at least have a chance of an outdoor life once their battery or intensively housed days are over. They are well worth serious consideration by the would-be poultry keeper, but may need a little extra care and management during their re-acclimatization period.

Used only to a small cage, ex-battery birds will probably be poorly feathered, unaccustomed to a nest box and floor litter in which they can scratch, and total strangers to a perch. Their toenails may also need clipping, although these should quickly shorten once the birds are out and about scratching. In an effort to stop feather-pecking (one or more birds pecking at another), some ex-battery hens will have had their top mandible (beak) trimmed, but there is nothing that can be done about this. To make life easier, however, it is probably best if you continue feeding dry layers' mash – which almost all commercial setups use: birds which have been de-beaked will find it far simpler to scoop the mash up with their lower mandible than to be forced to peck at pellets. Once the hens have settled, however, it should be possible gradually to introduce other feeds.

The Battery Hen Welfare Trust recommend introducing newly acquired birds to nest boxes by giving them access to a box on the floor, stating that 'even a cardboard box on its side with shavings and a bit of hay in will suffice until they are fit enough to use the proper facilities.' Because they have no concept of a nest box, birds will, in all probability, drop their eggs anywhere so you should encourage them to lay in the boxes provided by adding artificial or dummy eggs, bought from any agricultural suppliers. Encourage birds to roost by placing perches just a short distance above the shed floor and, if necessary, construct a ramp that will spur them to use the perches at night. Alternatively, lift them all onto the perches once dusk has fallen each evening – they will soon get the idea. When you let them out for the first time, do so a couple of hours before dusk, which will make them more likely to stay near to the house and return as darkness falls.

With a little care and management ex-battery hens (left) can be given a chance of an outdoor life (right).

COCK-CROWING COMPETITIONS

All cockerels crow, it's just that some do more so than others (a point always to be borne in mind when considering the purchase of a male bird – especially if neighbours are close by). Some produce a deep, dulcet tone; others, like the little game bantam cocks, give a high-pitched shriek. Over the generations, this easy ability to make a noise has been encouraged by some poultry fanciers and been developed into something of an art form.

Although sometimes held as part of modern village fetes and other fund-raising country activities, cock-crowing competitions were taken rather more seriously in America and parts of Europe during the early 1900s. Basically, the idea was for a time keeper to stand in front of each feathered contestant in order to note down the length of crow, the number of crows and the variation of notes in a given time – generally 15–30 minutes.

Cockerels considered worthy of being

entered for such contests were subject to strict training, part of which involved being kept in covered cages and taken into the open air each day before the cage covering was suddenly removed: thinking that morning had arrived quickly, the cocks would start crowing. Training went on for several weeks and each owner/trainer concocted his own special, and often secret, diet for his birds. Three weeks before a specific event, the feed was carefully dosed (with what can only be

An Old English Game bantam letting the world know of his presence.

imagined) and was said to have been made as 'stimulating and exciting as possible'.

In 1901, an anonymous writer described a crowing competition in the USA in the following way: 'The mode of operation is to place the cages containing the roosters in a long row, and then one bird will generally set the others crowing. A marker for each bird, appointed by the organizer of the show, has as his duty to note carefully the number of crows made by his bird. Each competitor puts up one dollar and the winner takes all.'

In Germany, the Bergse Kraaier (or Bergische Kraher) is thought to have been introduced by refugee monks to the village of Wuppertal, and then crossed with local breeds with the sole intention of producing a bird capable of winning local crowing competitions. Such competitions were not, however, confined to the US and Europe: the peasants of Imperial Russia developed the Yurlov Crower (see page 105) from ordinary village poultry which was bought from the Yurlov Fair in the hamlet of Izmalkovo. A contemporary writer remarked: 'It is interesting that these chickens have been breeding till now without any crossings with other cultural breeds but the cocks with strong voices are kept for breeder stock. Thanks to selecting the cocks with strong voice and therefore good health, peasants preserved the [loudest crowing] birds.'

Unlike France and other parts of Europe where a stop-watch was used, cock crowing in Russia was measured in 'quarters' – in quick succession by raising one of the fingers

of a straightened hand laid against the table edge (each quarter is approximately equal to a second). Training was non-existent; at the competition a cock bird was simply fed and its movements watched by the attendant crower fanciers, one of whom would suddenly slap one hand on his knee and shield his mouth with the other before making imitation crowing sounds. From all accounts this seems to have been sufficient to stir the competing cock bird into action.

Although nowadays merely a sound of the countryside, in years gone by a cockerel with a particularly fine voice may well have been entered into cock-crowing competitions.

HOUSING

Like any good home, the high start up costs of a chicken house are rewarded by a lifetime of comfort and productivity. You can choose from a wide range of commercially produced housing units, of which there are many at reasonable prices, or, alternatively, you can build your own, which can be tailor-made to fit a particular part of your garden or smallholding. Either way you will need to carefully consider the breed you have chosen, the number of birds you have, and the space availalbe before you make your decision.

SPACE REQUIREMENTS

The amount of space your birds need depends to some extent on the method of rearing you adopt. If they have access to a large outdoor space they will use the house only on wet days and for roosting or egg-laying. However, if they are kept intensively, the combined floor space of their house and run will need to be greater, as they will spend all their time there and will require sufficient room for feeding, roosting, nesting and exercising. Whatever system you choose, the basic premise is that you should give your flock enough room to ensure that the three Hs – health, happiness and hygiene – are all maintained.

If you intend to keep your chickens intensively a house measuring 1.5 x 2m (5 x 7ft) should be sufficient for six hens; if a run is attached, it will accommodate 12 birds. For these sorts of numbers a combined house and run area of 5sq m (6sq yd) per bird is adequate, although much will depend on whether you are keeping bantams, light large fowl or extremely heavy breeds such as Croad Langshans or Brahmas.

Remember also that unless your poultry-keeping operation is very small, one house and run is unlikely to be enough. There will be times when extra space will have to be found to accommodate a broody hen and her chicks, a 'resting' cockerel or a sick bird. If you plan to exhibit, you will also need a separate penning shed where birds can be kept clean and 'trained' to show themselves to their best advantage. In addition, a shed or some kind of food store may be required.

TYPES OF HOUSING

If you are keeping just a few birds a moveable combined house and run (sometimes known as an ark or fold unit) is probably your best option. You will, however, need sufficient space to be able to keep moving the run on to fresh ground before the area beneath it turns into a mud bath. The alternative to the moveable unit is a permanent house and run.

Moveable Housing

Some moveable units have exterior-opening nest boxes and, in a few cases, attached feeders and drinkers that can be accessed from the outside. The chickens can be kept confined but still have access to fresh grass and, if the unit is moved on a regular basis, there is little chance of a build-up of disease. The chickens can still be allowed free-range when you are around, but they are protected from predators and dogs when you are not.

The housing part of a fold unit usually takes up about a third of the overall length, with the remainder forming the run. The back of the run may be fully or partially covered to provide the birds with shade from the sun and shelter from the rain. Old-fashioned fold units, as once used by the commercial poultry industry, are still sometimes seen and are often equipped with wheels that make moving them easier. Most modern-day versions, however, merely have a carrying pole or extending handles at each corner. A strong floor is essential. The base of the unit needs to be thick enough to withstand moisture and being dragged on to fresh ground regularly.

CUSTOMIZE YOUR CHICKEN HOUSE

You can customize a bland wooden house and create something that will blend into your own environment. You could paint it your favourite colour, write your chickens' names above the door or maybe train plants to grow against it so that it blends in with your garden. Remember to allow plenty of ventilation near the top of the house and ensure that it is draught-free – chickens really don't like the wind and the rain blowing into their house.

Above: The best of all worlds, this fixed unit provides shade in the summer and shelter in wet weather, and allows the birds access to the garden when convenient.

Right: A moveable fold unit is the perfect solution for keeping chickens in small gardens. The carry poles at either end enable it to be easily moved, and this should be done every few days to give the chickens access to fresh grass while allowing the previous patch to recover.

If you have only a trio of large birds or half a dozen bantams, an oversized rabbit-hutch type unit is ideal. One-third comprises a wooden-floored roosting and nesting section, while the remaining two-thirds has no floor, giving the occupants access to fresh grass. As only the front of the run section is wired, this type of housing is totally wind- and waterproof and so would be worth considering if you intend to keep light-coloured exhibition birds. Some designs also permit the whole roof to slide back on runners, which makes routine maintenance very easy indeed.

For some of the true bantams or miniaturized breeds, an actual rabbit hutch might make a good home. The birds would, however, need access to a small exercise pen or should periodically be given free-range to roam the garden.

Permanent Housing

The Victorians often made a feature of their henhouses in the same way that dovecotes evolved to become ornamental. Poultry houses usually took the form of a lean-to along the side of a walled garden or stable yard, and consisted of a shed and run, with the roof of the shed continuing over the run section. Separate doors allowed access to either the house or the run, the floor of which was normally constructed of compounded earth.

If you have the space and a suitable wall that is not too close to your own house or that of your neighbour, you might consider a variation on this theme. If you replaced the floor covering with sand, such a set-up would be ideal for some of the heavily leg- and feet-feathered breeds. The covered run would

also be perfect for light- or white-coloured exhibition birds, which become brassy if exposed to the sun or inclement weather.

A large conventional henhouse can often be partitioned to allow several pens of birds to be kept under the same roof. The only disadvantage to this is that a narrow corridor needs to be constructed through the interior of the shed, so that the daily chores can be carried out from the inside. This removes the possibility of the chickens having access to two runs, one on either side of the building, which they can use alternatively.

Instead of buying a commercially available henhouse, you could adapt an existing outbuilding. It will need to be light, airy, well ventilated and predator-proof, and the door

should open and close easily, as you will need to use it several times each day for access. Brush away any cobwebs on the inside of the shed, wash the windows and, if necessary, paint the walls to make it brighter.

It is important that you take the possibility of hot summers into account if you are planning to use an existing structure to house your birds. Old stone buildings are well insulated from the sun, but more modern sheds may become stifling in the heat of the day.

A bright, healthy-looking comb is normally a sign that a hen is in full lay.

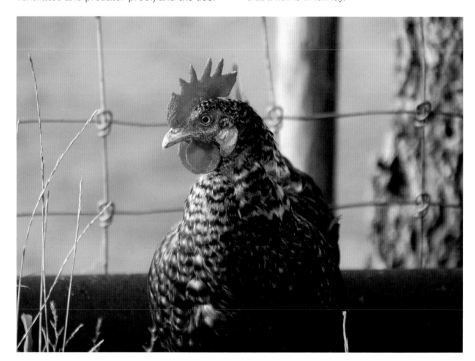

STRUCTURAL CONSIDERATIONS

Whether you are designing your own henhouse or purchasing a ready-made unit, it is important to bear the requirements of your chosen breed in mind. If you intend to keep Yokohamas, for example, the house needs to be tall enough to contain perches fixed at such a height that the birds' flowing tails will not be damaged when they are at roost. Likewise, a house with a pop-hole only 30cm (12in) square might suit bantams, but it will make it impossible for a large Buff Orpington to exit or enter.

Nor should houses be bought for the convenience of the chickens alone. If you have a bad back, for instance, you will not enjoy having to get down on your hands and knees every time you need to feed your chickens, collect their eggs or clean out the roosting compartment of a small ark unit. For the same reason, permanent housing must be at least 1.5m (5ft) high to allow easy access.

The floor of a permanent henhouse must be sufficiently strong and adequately braced to prevent springiness as you walk on it.

Raised houses must also be equipped with a gangplank leading up to the pop-hole in order to prevent possible injuries to the chickens' legs and feet as they come and go.

WINDOWS AND VENTILATION

Good ventilation in the henhouse is vital in preventing respiratory diseases, and can be provided by natural means such as windows and/or a protected open ridge running along the roof apex. It must be possible to open the windows, but if there is any danger that predators or vermin can get in through them, they should be covered with small-mesh wire netting on the interior of the building. Approximately one-fifth of the available wall space should be given over to windows. Ventilation is important all year round, but make sure your chickens are not exposed to draughts in cold weather.

NEST BOXES

The henhouse should contain some simple nest boxes, ideally one for every three laying birds you intend to keep. Each box should measure 30cm (12in) square and be around 35cm (14in) high. They may be made in tiers, in which case an alighting perch should be attached to the front of each box.

The boxes should be placed just off the ground and in the darkest part of the house (usually under the windows), as chickens prefer a quiet, dim place in which to lay their eggs. Straw or hay makes a good liner, but it will need to be cleaned out regularly and dusted with louse powder. Remove any broken eggs immediately to discourage egg-eating, a habit that, once formed, is very hard to stop (see page 142). Occasionally, new birds try to roost inside nest boxes, but this must be discouraged or foul nests and dirty eggs will result. When the chickens are first placed in the henhouse, prevent them from roosting in the nest boxes at night by covering up the fronts of the boxes with a board or a curtain of sacking. For the same reason, cover the top of each box with a steeply sloping board.

If you want to get into serious egg production, consider using trap nests, as they are the surest way of telling how well a bird is laying. The trap closes when a hen enters the nest to lay, so she cannot get out until she is let out. Combined with the use of leg rings or wing tags to identify the individual birds, trap nests allow you to keep a record of all the eggs each hen lays. The system demands that the house is visited every two to three hours, but if you have the time the results can be very interesting.

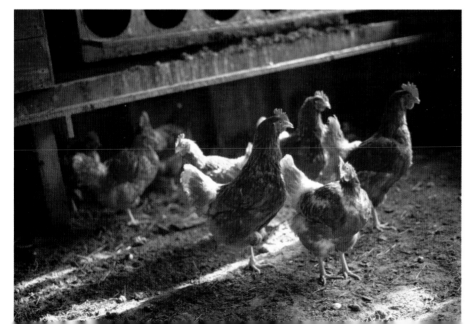

Chickens like warmth but not too much heat: make sure they always have shade available.

TYPES OF BEDDING

Whatever materials have been used to build the henhouse floor, a covering of litter is essential. It serves three main purposes: to absorb the birds' droppings; to provide ground-level insulation; and to give the birds plenty of exercise and interest. Types of bedding include:

Wood shavings are the most common form of litter, available in compressed bales from suppliers of horse feeds and equipment. If you manage to source your own local supply from a firm making wood-based products, check that the shavings are safe to use (softwoods are best) and that they have not been treated with toxic or irritant chemicals. Because sawdust is fine and is made even more so by the birds' constant scratching, it may cause respiratory problems and so should be avoided.

•

Shredded newspaper is available in bales from most agricultural stockists. This makes perfect bedding material as it is highly absorbent, contains no parasites and can be burnt or added to the compost heap after use. The only disadvantage of newspaper is that it is so lightweight – on a windy day the surrounding area can end up looking as though a ticker-tape parade has just taken place.

•

Straw is probably even easier to obtain than shredded newspaper in rural areas, as most farmers will be able to sell you a bale or two after harvest. Wheat straw is best, as it is most durable, but oat straw is a good second choice. Although chopped straw is more absorbent and is preferred by chickens, it can, like sawdust, become dusty and affect the birds' respiratory system. You should never use hay, as it mats together into a carpet and has also been known to make birds crop-bound (see page 148). It can, however, be used in nest boxes or when forming a comfortable base for a broody hen to hatch fertile eggs.

•

Leaves are collected in the autumn. Birds will happily scratch around in these. At one time, peat was favoured by poultry keepers, but it is very dusty unless kept damp and collecting it is not environmentally friendly.

•

Silver or sharp sand may prove to be the best floor covering if you keep a feather-legged breed. Ask your intended stock supplier for a recommendation.

Whatever litter you choose, make sure that it is not mouldy or damp. Mould can cause serious respiratory diseases, while wet litter is a perfect breeding ground for the *E. coli* bacterium. The litter base should be about 12–15cm (5–6in) deep and needs to be raked over daily to remove any matted portions, such as under the perches and around water drinkers. In addition, the litter should be renewed completely every five or six months (see page 118).

PERCHES

One or two perches running the length of the henhouse will give your chickens or bantams somewhere to roost at night. If you need to install more than one perch, be sure to keep them at the same height or the birds will all try to roost on the higher one.

For bantams, the perch should be about 5cm (2in) wide and 3cm (1in) thick, while large birds require something slightly more substantial – around 8cm (3in) wide and perhaps 5cm (2in) thick. In all cases, the top edges must be rounded off. Always try to use planed wood for perches and never use any wood that has bark attached: rough-sawn timber can cause splinters, while a covering of bark provides an ideal home for parasites such as red mite (see page 149). Both types of wood are also difficult to scrape clean.

At least 20cm (8in) of perching space

Measure the pop-hole is large enough for the breed you have chosen to go in and out comfortably.

Fine-mesh wire at ground level should keep your birds in and predators out.

should be allocated to each bird, and if a single perch is being used in a large poultry house it should be supported every 1.8m (6ft) of its length. The height at which the perches are fixed is also very important. Unless you are keeping long-tailed breeds (which require a higher roosting point), around 60cm (2ft) off the floor is about as high as you should go, although for some of the heavier breeds such as Croad Langshans, 30cm (1ft) is perfect. Perches should not be too high – even for the lighter, flightier breeds – because the pads of the birds' feet can be injured by constant hard landings each morning. If you use a droppings board, set it at around the heights recommended above for perches and then construct the perch about 15cm (6in) above.

If you are housing exhibition birds, it is particularly important that their tails do not become damaged by constant brushing against the interior wall. The distance between the wall and the intended perch position should take this factor into account.

Perches must never be attached permanently to the walls of the house. Instead, they should be fitted into slotted grooves, hooked into brackets or built as a free-standing frame. In this way, they can be removed for regular cleaning and disinfecting, and they can also be taken out altogether if the house is required for a broody hen or for young stock.

RUNS AND FENCING

The fencing used to form a run around permanent poultry buildings should be around 2m (6ft 6in) high, not only to contain flighty birds but, more importantly, to keep predators out. In practice, the chicken wire

most commonly used for this type of fencing has a 19-gauge thickness and 5cm (2in) mesh, and measures 1.8m (6ft) across the roll. The fence posts should be around 4m (12ft) apart. About 30cm (1ft) of the fencing wire must be dug into the ground to prevent pests or predators from scratching their way in or the chickens from pushing their way out, and unless the top of the pen is covered with a nylon net (see below), brackets of about 1m (3ft) should be attached to the top of each fence post on the outside. Three strands of barbed wire (or electric wire, for which you will need a fencing unit) can then be affixed to each bracket quite tightly, and this will prevent most predators from climbing into the pen.

A nylon net can be used to form a roof over the run, propped up with stakes to keep it taut, although upturned plastic plant pots will need to be placed on top of each stake or they will eventually rub a hole in the mesh.

When building a gate to the run, make sure that the method used to fasten it is secure. A central bolt often seems safe, but a

determined fox or dog pushing and scratching at the base of such a gate could force an entry. Top and bottom latches are the better choice. The gateway must also be wide enough to allow for the passage of a wheelbarrow or any broody coops and runs that may be needed in the future.

Where different breeding pens of birds share the same dividing fence, it is imperative that some sort of screen is in place to prevent the cock birds from seeing each other. Without it, they will spend more time trying to fight one another than they will mating with their hens, and some serious injuries could result. The easiest and least expensive way of forming this partition is by using one of the many windbreak materials sold by garden centres and horticultural suppliers. Sheets of corrugated metal are another long-lasting option and, provided that they are painted to blend in with the surroundings, will not look obtrusive. If finances allow it or if you know of a low-cost source of planking, solid wood partitions can also be used, but they will need to be at least 60cm (2ft) high for bantams and twice that for large fowl.

If your birds are free-range, electric netting can be used to create temporary runs as a means of protection. The horizontal wires of this netting are electrified and the thick black strand at the base acts as an earth.

If you have a small permanent house with an integral run, a concrete base over which gravel, sand or bark chippings can be laid provides a hygienic base. Birds kept in such conditions will remain happy and healthy, so long as they are given a regular supply of greenstuffs in addition to their usual diet.

POSITIONING THE HOUSE AND RUN

Chickens can tolerate cold weather but dislike draughts; they love warmth but are uncomfortable in extreme heat; and they delight in pecking through friable soil in search of insects but hate getting their feet wet. The ideal spot for a run therefore needs to be well drained and surrounded by some form of windbreak, with the front of the henhouse positioned to face the morning sun but protected from its hottest rays at lunchtime and in the early afternoon. At least part of the run should provide some shade at all times of the day.

Also bear in mind the run's location in relation to your own house and those of your neighbours, as even the best-maintained sites may encourage flies and a few unpleasant smells during the summer months. While you will enjoy watching the antics of your birds as they rush importantly about their business, your neighbours may not, so screening the run with trelliswork, hazel hurdles or other structures over which plants can be grown

may well be appreciated. However, do be careful not to erect anything too close to the chicken run that could be used by predators as a means of gaining entry.

Do not site a permanent run on heavy ground with a high water table. If there is no alternative but to place it in a low-lying area, grade the site so that surface water runs away from the building. It is also a good idea to surround the house with a 2m (6ft 6in) apron of gravel or walkway so that it does not become a mud bath in winter. Fix guttering to large houses so that rain falling on the roof drains away, either into a water butt or a piped trench.

Large houses should be raised off the ground so that rats and other vermin cannot take up residence beneath them. The space under a raised house will also provide extra room for your chickens and is an ideal place for them to dust.

Where chickens are confined to a wire run, they will soon make their own dust bath, but

a covered structure is very easy to build and will be much appreciated. All that is required is a small shelter consisting of four posts and a sloping roof about 1m (3ft) high. Turn the ground over and leave the chickens to do the rest. You might like to add some fine soil, wood ash or sand to the mixture, but the most important thing is that the area remains dry.

CLEANING ROUTINES

Inside the henhouse, it pays to gather droppings regularly – preferably each day but certainly at least once a week. Make it part of your daily routine to remove the debris from the droppings boards below the perches and to replace this with a light covering of fresh wood shavings. About once a month in the summer and every two months in winter, nest boxes need to be cleaned out and the bedding replaced.

In addition to the daily removal of droppings, all the floor litter in permanent houses must be removed at least once a

A moveable house makes it practical to give your birds regular access to fresh grass. If the top of this one had been extended to provide handles, it would have been easier to move.

year (preferably twice) and the interior thoroughly disinfected. Try to pick a sunny day for this job and lock the birds out of the shed. Remove as many of the fixtures as possible, including perches, droppings boards and nest boxes, and disinfect these and the interior with a virucide. A knapsack sprayer is more effective than a bucket and brush, as the higher pressure created ensures that the killing agents reach all the nooks and crannies where parasites love to hide. A steam-cleaning pressure washer is perhaps the best means of application; small units can be bought quite inexpensively.

At the same time as you disinfect the house, remove all covering litter from attached runs that have a concrete or compacted earth base. If you do not, there is a danger that parasites and bacteria could immediately be reintroduced to the house via the chickens' feet. If gravel, sand or bark chippings are used as the litter, it may be possible to rake or riddle the droppings and rubbish from the material, then disinfect it and re-use some of it, thereby cutting down on the cost of buying replacement litter.

Being smaller, moveable fold units present a less daunting prospect when it comes to cleaning, so it is not too onerous a task to clear them out and disinfect them every six weeks or so. Even then, it is all too easy to let the six weeks become ten, so to avoid slippage keep a note of when the next 'service' is due.

Feeders and drinkers should be kept clean at all times. Attach a small, stiff-bristled brush to your water bucket or watering can so that you can brush and rinse out the drinkers before refilling them with fresh water each morning – this takes no time at all. Feeders will not get too dirty if dry pellets or mash are used, but they will receive their

share of droppings: clean them out as part of your daily routine. Every few weeks, drop both drinkers and feeders into a proprietary cleaning solution and give them a thorough scrubbing.

Do not forget to give incubators, coops, brooders and other equipment not in daily use an annual spring clean. Incubators and brooders should preferably be fumigated both before and after use. You will find fumigation products at an agricultural stockist or advertised in poultry magazines.

MAINTAINING THE HOUSE AND RUN

As with most equipment, the regular maintenance of poultry houses and runs will ensure that they last for years. By replacing a loose board, re-stapling a piece of sagging wire netting, oiling hinges and regularly treating all exposed woodwork with a preservative, you can extend the life of these components almost indefinitely.

Obviously, any wood preservatives must not contain chemical compounds that are likely to be harmful to the birds, so read the instructions on the packaging carefully before making a purchase. With the majority of modern products it is possible to treat the henhouse both inside and out, and then allow the chickens to return home just as soon as the wood is dry.

In addition to checking the henhouse itself, periodically inspect the perimeter of your runs for breaks in the fencing through which a predator could gain access. Although good-quality brands of chicken wire will last for up to 20 years, some inferior ones are made of lower-gauge wire and are not galvanized quite as thoroughly. It is a good idea to keep the fence free of vegetation so that any holes are more easily noticed.

An access ramp prevents birds having to jump down from a raised house, and possibly injuring their legs and feet.

BUYING

So you've decided on your favourite breed and type of chicken; now you just need to find a high-quality supplier. Poultry shows and magazines can be great sources of information, but word-of-mouth recommendations are generally the best way to obtain your stock. Bear in mind that breeders are unlikely to sell you their very best birds, as they will want to keep these to maintain their own strains.

BUYING AND SOURCING BIRDS

In addition to publications aimed specifically at the poultry keeper, the classified section of your local newspaper may contain advertisements for poultry. Birds are quite often sold as trios (one cock and two hens), but in some countries a numerical code is used to describe what exactly is on offer. Such an advertisement may include the name of the breed, followed by '1–2', denoting one cock and two hens; '1–0', telling you that a single cockerel is on offer; or '0–3', indicating that three hens or pullets are looking for a new home. Do not be confused by numbers like '06' at the end of the advert, as these merely indicate the year in which the chickens were bred rather than the quantity on sale.

If you have set your heart on a particular breed you may have to travel some distance to find it, but with effort and perseverance you will eventually source exactly what you want.

USEFUL ORGANIZATIONS

From your initial enquiries about breeders, you should also be able to find out the whereabouts of your local poultry club.

The addresses of clubs and societies devoted to specific breeds can be found in specialist poultry-keeping magazines or through the national poultry clubs. In the UK, these are the Poultry Club of Great Britain and the Rare Poultry Society (see box), while in the US, the American Bantam Association and American Poultry Association are on hand to answer questions. The breed clubs generally hold shows and exhibitions several times a year, and these are always good venues at which to make specific enquiries about purchasing birds and to gain general advice. Sometimes they also include selling classes, which are perhaps the best way to obtain your desired stock.

Most of the clubs have websites, and in addition to these there are many other poultry-oriented websites that offer chickens for sale, information on breeds, and noticeboard links where you can post your requests or simply communicate with like-minded enthusiasts.

THE AGE TO BUY YOUR BIRDS

If your chosen breed proves difficult to find, or if your finances will not permit the purchase of adult pure-bred birds, you might consider purchasing a small pen of bantams to use as surrogate mothers to

A clear, bright, well-pigmented eye and brightly coloured comb and wattles are good indications of a healthy bird.

'Selling' classes, which feature in some poultry shows, can be a good way of obtaining quality stock.

USEFUL ADDRESSES

Poultry Club of Great Britain
Keeper's Cottage
40 Benvarden Road
Dervock
Ballymoney
Co. Antrim
BT53 6NN
www.poultryclub.org

Rare Breeds Survival Trust
National Agricultural Centre
Stoneleigh Park
Warwickshire
CV8 2LZ
www.rbst.org.uk

Rare Poultry Society
www.rarepoultrysociety.co.uk

Utility Poultry Breeders
Association
Morville Heath
Bridgenorth
Shropshire
www.utilitypoultry.co.uk

American Bantam Association
PO Box 127
Augusta, NJ 07822
www.bantamclub.com

American Poultry Association
PO Box 306
Burgettstown, PA 15021
www.amerpoultryassn.com

hatch fertile eggs instead. Such birds must be the right sort: heavy in appearance, placid and reasonable egg-layers. As they go broody (which they certainly will), you can set fertile eggs bought from a known breeder beneath them to hatch. Such eggs can be bought and transported over long distances much more cheaply than mature stock. It might sound risky to entrust such a fragile parcel to a postal or private delivery service but, provided the eggs are carefully packed, there are rarely any breakages.

Another option is to buy day-old chicks and rear them using bantams in the same way.

Many of the larger poultry breeders hatch their eggs by artificial incubation and are quite happy to sell chicks to chicken fanciers. Where possible, you should collect the chicks in person but, like eggs, day-olds do not seem to suffer unduly by being transported by rail or road. The methods of introducing eggs or day-old chicks to a broody hen are described in Breeding Chickens, page 132.

Purchasing adult birds obviously has its advantages. Not least, you do not have the frustration of rearing young chicks to maturity, only to find that they are either nearly all cockerels and therefore unable to supply eggs, or, if showing is your ultimate goal, sub-quality specimens that do not quite conform to the breed standards. Birds bought at point-of-lay or beyond can be placed in the chicken shed and run immediately, and do not require any intermediate housing or extra care. And the maintenance costs of adult birds are lower as there is no need to provide heating, chick crumbs or growers' pellets.

A pullet is a female bird that is in its first year and has yet to commence laying; the word 'hen' generally refers to an older female that has completed her first season of laying. From the point of view of buying birds, pullets are a better choice than hens as you will obviously gain more eggs from them over the course of their lifetime. Having said that, if eggs are your main objective there is no reason why you should not consider buying hens that have been discarded by commercial set-ups after their first season, as these birds are usually inexpensive and still have plenty of laying potential ahead of them.

When contacting breeders initially, it is important to state whether you want pullets or hens. Likewise, when selecting a male bird, specify either a cockerel (under 18 months) or an older cock bird.

WHEN TO BUY

In addition to selling surplus birds at shows, poultry clubs often organize poultry sales and auctions. When these are held in the early spring, you can be sure that the majority of birds on offer will be around a year old and in peak condition for breeding or just ready for a long life of egg-laying. Sales held in the autumn may contain young birds bred in the same year or older birds that, while they might be past their prime for showing purposes, are nevertheless still capable of producing eggs or show-quality offspring in the next rearing season.

Generally speaking, at autumn sales it makes sense to buy stock that has been hatched in the spring of the same year. By the time they are sold, such birds can be kept as adults and should lay, if somewhat sporadically (unless they are a commercial breed), throughout the winter months before really getting into their stride the following spring. Show birds of this age will also be at their peak for breeding and exhibition purposes the following year.

If you are not buying at spring and autumn sales, just choose the best birds as and when they are on offer. However, remember that chicks or young birds always do better and grow more vigorously during the summer months, when they have access to fresh grass and more sun on their backs.

Chicks bought day-old need to be reared under a heat lamp or a foster mother.

WHAT TO LOOK FOR

It is useful to be able to recognize the signs of a healthy bird from its overall appearance. First, the comb and wattles must be bright and waxy. There are a few exceptions to this rule, such as when the breed has an unusually coloured comb (for example, the Silkie), or when a hen has been brooding chicks or is in moult and not laying. Second, the eyes should be clear and bright and the plumage shiny and full. During the summer, you may notice broken feathers on the back and neck of females in a breeding pen, but this will probably have been caused by the mating cock bird.

Handle any bird you consider buying, as even the breeds classified as light should feel fleshy and well muscled. Check the breastbone – it should be reasonably well covered with flesh and certainly not sharp like a knife blade. Inspect the vent (anal area) for signs of diarrhoea and for lice and mites. The underpart of the wings nearest the body is another area favoured by parasites, so lift up the wings, gently brush the fine feathers in the opposite direction to their growth and look carefully for movement – although tiny, fleas, lice and mites are all visible to the naked eye.

Finally, spend some time watching the flock: they should all be busy scratching around, dusting in the earth, feeding or preening. Avoid buying birds from a pen where even just one individual is moping around or is isolated from the rest of the batch, as this could be a sign of disease and at the very least indicates that all is not well. Also look around the pen to ensure the faeces are firm, well formed and, in part, white in colour. Slimy, watery green or yellowish diarrhoea-like droppings are another indicator of ill health.

GETTING YOUR BIRDS HOME

Before setting out to collect your chickens, make sure that the henhouse and run are ready for their arrival and that food and water are in place. It pays to ask the breeder in advance what type of feedstuffs your new birds are used to and, if possible, to purchase your food from the same manufacturer. Alternatively, ask the breeder if you can take home some food so that you can make the changeover to a new supply gradually.

If you don't have a proper poultry crate or basket for transporting the birds arm yourself with a sturdy cardboard box that has plenty of air holes and a decent lid. Construct hinges and fasteners by threading wire or string through small holes drilled in the box and lid. Cover the base with hay or straw.

Once home, keep the chickens confined to their house for a few hours so that they can become accustomed to their new surroundings. When they appear settled (after a few hours in the day or overnight), lift the pop-hole and let them explore the run area. Don't push them out: it is better to let them make their own way. If the house has no confining run, leave the birds shut in for 24 hours before giving them free-range.

Make every effort to get acquainted with your chickens straight away. Always move slowly and spend some time crouching at the feeder, so that hunger and interest will ensure they approach you. Before long, they will quite literally be eating out of your hand.

Transport birds in a strong container that is not too big for them, otherwise they will slide about in transit.

FEEDING

If you feed your birds regularly, and with the correct food, you can expect eggs, meat for the table, or a good crop of chicks. But feed them badly and they will fail to produce anything at all, as the little nutrients they receive will be taken up with keeping their body functioning correctly. Fortunately, feeding chickens is simple and relatively inexpensive. They are industrious creatures and will find insects, seeds, grubs and fresh grass themselves in their run, especially if it is a folding unit that can be moved regularly. On top of this, they will need two meals a day, the contents of which can be purchased from any agricultural supplier or pet shop.

When selecting feeding equipment, allow at least 10cm (4in) of space per bird.

WHAT CHICKENS EAT

Chickens are omnivores and, provided they are given the opportunity to do so, will feed on seeds, herbs and leaves, grubs, insects and even small mammals such as mice. All fowl are foragers, with an evolutionary instinct to range and search for food. They have excellent full-colour vision and a highly developed sense of hearing that, thousands of years ago, enabled them to keep track of the rest of the flock while ranging over vast areas of dense foliage in search of food. Even today's highly domesticated breeds have the same desire to hunt and, in the case of poultry, to scratch in their search for food.

Domestic chickens are typically fed commercially prepared feeds that include a protein source, vitamins, minerals and fibre. This usually takes the form of crumbs, pellets or mash, although most types of poultry (there are certain exceptions, such as laying birds) also benefit from the addition of mixed corn to their diet.

Poultry have differing nutritional requirements depending on their age. Young chickens and bantams, for example, need much more in the way of protein when they are growing fast than when they are older and begin to grow more slowly – between, say, 10 and 18 weeks.

THE DIGESTIVE SYSTEM

Chickens have a very simple digestive system. Food is taken via the gullet (oesophagus) into the crop, where, by the end of the day, it will often show as a bulge in the chest, just above the breast and below the throat.

The food is softened in the crop – mash and pellets break down almost immediately, whereas it can take several hours for cereals to do the same. The food then passes through a glandular stomach (proventriculus), which predigests it, and into the gizzard, where it is ground down by muscular action. As birds have no teeth, they would find it nearly impossible to make use of much of their food, especially wholegrains, without the aid of insoluble grit in their gizzard (see page 129).

NUTRITIONAL REQUIREMENTS

Birds need protein for growth, for tissue repair and for developing immunity against disease. Fats and carbohydrates are used to provide heat and energy, any surplus being stored in the body as fat, and fibre is required in small amounts to keep the bowels working. Finally, vitamins and minerals are needed to maintain health, while the latter are also used in bone and eggshell formation.

In the wild, a bird's food contains all the fats, carbohydrates and plant and animal proteins it requires, plus sufficient levels of naturally occurring vitamins and minerals. However, as domesticated birds, chickens no longer have the freedom to make up their own nutritional package, so it is important that you offer them a diet that is well balanced, providing all their needs at each stage of their life.

BROODY HENS

A broody hen eats very little when she is sitting, and requires little more than a simple diet of mixed corn, with plenty of maize to maintain body mass and warmth, along with fresh drinking water. No other specific nutrients are required. If fed grain she will also produce firmer droppings, making her less likely to foul her nest.

CHICKS

Even if you have a home-produced or organic feeding programme in mind, do not be tempted to rear young chicks on household scraps. Without the correct protein levels at this stage, they are likely to die or, at best, will end up undernourished. Proprietary chick crumbs are essential, because they contain not only all the nutritional requirements, but also some necessary medicinal additives (to combat coccidiosis, for example – see page 151).

In most cases, chick crumbs can be phased out at around three or four weeks of age, but some inbred strains or more delicate true bantams may require crumbs for twice as long, as they are often less efficient at absorbing all the necessary nutrients from their food.

GROWERS

Proprietary growers' rations are the most sensible option for growing birds. They have a lower protein and vitamin content than chick foods. Pullets must not be overfed or given too much protein, or they will grow too quickly while still being immature internally. This will cause various egg-laying problems and possibly a partial moult at point-of-lay (see page 146).

LAYERS

Buy a good balanced layers' ration that provides around 17 per cent protein, and feed it either as a coarse-ground mash or as pellets. Grain or other low-protein foods should not be given with layers' feeds, as their addition will unbalance the manufacturer's carefully evolved formula, although in practice this appears to do little or no harm to the small, non-commercial flock.

Your birds will appreciate a few treats such as sultanas, which will also help when you are taming new stock.

TABLE BIRDS

If you are raising surplus cockerels for the table, household scraps can be an invaluable addition to their diet when combined with a high-protein proprietary fattening meal, and will certainly help in bulking them up. A bird fed in this way will be a great deal tastier than the majority of commercially produced chickens, and its varied diet (especially if maize is given) will make the carcass virtually self-basting. With this method of feeding, a bird is ready not after a certain number of weeks, but when it has put on enough bulk – pick it up, assess its weight and feel its breast (see page 146). Cockerels as young as ten weeks can make magnificent eating; in the catering trade these are known as poussins and command a high price.

BREEDING BIRDS

Birds that are used for breeding need a plentiful supply of protein and trace elements in their diet or their offspring may suffer from deficiency-related diseases.

Breeders' rations are, as their name suggests, suitable for inclusion in the diet of birds you are intending to breed next year's stock from and should be fed from mid-winter onwards.

Maintenance rations can also be fed to breeding birds, and in fact can be used during any period of a bird's life when its body is just ticking over, such as when it is in moult.

You may wish to consider buying breeders' and maintenance rations produced for game birds, as these are some of the best feeds available.

COMMERCIAL FEEDS

Assessing and combining all the required ingredients of a chicken's diet is a fine art, but fortunately there is no need to produce a perfect home-made mix now that balanced commercial foods, including free-range and organic brands, are available. Whatever your choice, buy high-quality foodstuffs from reputable firms.

Ornamental fowl pellets are available from some of the more specialist food suppliers. They are especially valuable if you want your birds to be in prime condition immediately prior to the show season. The pellets are also smaller than regular pellets, making them ideal for bantams.

PELLETS OR MASH?

While pellets may be the best option in terms of ease of use and reducing wastage, some experienced poultry keepers think dry mash is a better form of feed. Because birds can consume their entire day's ration of pellets in half an hour, they soon become bored, especially in winter. By contrast, it takes them about three hours of continuous feeding to eat the equivalent amount of mash. If you are keeping birds for exhibition, however, pellets are your only option as mash is too messy – show birds, especially the bearded or muffed varieties, will end up with food-encrusted feathers around their beak.

Intensively reared chickens bought from commercial farms after their first season of egg-laying will almost certainly have been fed on mash. If you wish to change their diet, it is possible to wean them on to pellets and mixed grain. However, to build up their stamina

This circular hopper has the advantage of being virtually waterproof, so the food is less likely to get damp.

when they first arrive, give them a warm mash mixed with such 'treats' as cooked vegetable peelings, cooked rice or crushed cornflakes. This will soon turn a poorly feathered, lacklustre hen into a blooming individual.

If you are using mash, stir it well before feeding. Studies have found that mash ingredients such as ground limestone (included as a source of calcium) tend to sift through the food and settle at the bottom of the bag. Birds that are given food from the top of the bag may, therefore, miss out on some of these vital components.

If for any reason you wish to change the type of food you give, do so gradually, over a period of seven to ten days. Any changeover should take roughly the following pattern: mix in around a quarter volume of the new food to three-quarters of the old for two or three days; mix half and half for the same amount of time; give three-quarters of the new food mixed with a quarter of the old for another two or three days; then finally feed the new ration entirely.

CEREAL FEEDS

It is pointless trying to give a combined feed of pellets and cereal at the same time, as most birds will eat the wheat first and leave the rest. As cereal is harder to digest, it is best given on its own as the afternoon or early evening feed, thus ensuring that the birds go to roost with a full crop. The afternoon feed is also one occasion where an exception can be made to the rule of feeding chickens indoors. If the corn is scattered around the outdoor run among dead leaves or other suitable scratching material, they will gain a great deal of amusement and exercise.

The best mix for a cereal feed is wheat and maize, around one-quarter maize to three-quarters wheat, though a little more maize can be fed during the winter months, as it helps the bird to maintain body heat. Too much maize in a laying bird's diet is bad for its health, as fatty deposits can build up around the ovaries, but a diet high in maize will fatten table birds more quickly and give their flesh that 'corn-fed' colour. Split maize is a very useful tool in enhancing the colour of yellow-legged breeds intended for showing: including it in their diet will often sharpen up faded legs. In addition, it improves the colour of egg yolks.

You can buy bags of ready-mixed grain but you may find that they contain barley, which chickens and bantams dislike and often reject in favour of other grains. It is therefore generally more effective, and less expensive, to buy wheat and mix in the maize yourself.

Sunflower seeds, linseed and other even smaller seeds intended for feeding to wild birds can be bought to help keep domestic poultry amused, although they should be given only as a treat rather than forming part of the daily cereal ration. It is more economical to buy such items in bulk rather than in the smaller bags sometimes seen in pet shops and supermarkets, but this is worthwhile only if you use them up before they reach their expiry date.

A mixture of wheat and whole or split maize makes a healthy afternoon feed.

Your birds should always have an adequate supply of clean, fresh water.

GREENSTUFFS

Chickens need greenstuffs for the minerals and vitamins they contain, and in the case of free-range birds most of their requirements can be obtained naturally. In the early part of the year, grass that has been managed correctly will be high in chlorophyll which, like split maize, improves the colour tone of yellow-legged chickens. Birds kept in runs or paddocks will also manage to find greenstuffs in the form of grass or weeds, but those in grassless runs will have to be fed a regular supply. A few turfs of short grass can be thrown into the run at intervals, but it is far better to hang fresh leaves off the ground or to put them in racks every day. Whatever method is chosen, any leftovers must be removed each night and placed on the garden's compost heap. In addition to greenstuffs, windfall apples and any sweetcorn that has gone starchy and is too hard for use in the kitchen will also be readily devoured.

If you have space in your garden, you may consider growing crops especially for feeding to your poultry. Useful feeds include lucerne, vetches, green cereals, rape, kale, millet and clover. The birds will also appreciate kitchen vegetables that have gone to seed. Always try to give them the whole plant, as they will benefit from the grubs and minerals in the soil surrounding the roots as well as from the leaves and stems.

Weeds gathered in the countryside, provided they are free of agricultural or car-exhaust chemicals, can be another good addition to the diet. Young chicks in particular appreciate chopped dandelion leaves, plantains, shepherd's purse, watercress and, appropriately enough, chickweed.

WATER

Never let your birds go without clean, fresh water. How much they drink in a day depends on the time of year, the type of food being given and the size of the bird. As with feed, you must be the best judge of how much your flock drinks, something that can be estimated only by careful observation on your part.

HOUSEHOLD SCRAPS

For maximum egg production, household scraps should not form more than 20 per cent of your hens' daily food intake. If you are keen to utilize food scraps, you can boil them up and add them to mash, but make sure that the resultant mixture is moist but crumbly, not sloppy: it should break up easily when a mixing spoon is put through it. Although this feed (called wet mash) is favoured by some poultry keepers, it is very time-consuming to prepare and you can put out only as much as the birds can eat in about half an hour or it will become sour and unpalatable.

Chickens enjoy stale brown bread (white bread is as unhealthy for poultry as it is for humans) soaked in milk, cheese, rice, pasta and other cereal-based scraps. Do not feed strong-tasting, mouldy or otherwise contaminated food, as this will, at the very least, taint the eggs and can cause digestive problems and even death. The discarded outer leaves of green vegetables can be given to chickens as part of their daily diet of greenstuffs.

GRIT

Flint grit is essential in allowing birds to digest food once it has entered the gizzard (see page 125). Because stones on the ground are usually rounded, they have little effect on grinding food into the necessary consistency, so flint grit should be given separately. It can be purchased from any agricultural supplier – expect each bird to consume approximately 30g (1oz) per month. Rather than putting the grit in a separate trough, some poultry fanciers put a thin layer in the nest boxes in place of hay or straw so that the hens pick up all their requirements when laying.

Grit is also a means of supplying calcium and, although this is often included in manufactured feedstuffs, it can be given separately as oyster shell or ground limestone. Many chicken keepers believe that birds fed good-quality layers' ration should not be given extra oyster shell because this will upset the calcium:phosphorus ratio, resulting in thin-shelled eggs, brittle bones and, in some cases, feather loss.

VITAMIN AND MINERAL SUPPLEMENTS

While vitamin and mineral supplements are available commercially, they are not normally needed because manufactured pellets and mash always contain the usual daily requirements. Birds that have access to grass will usually find sufficient minerals and vitamins as they scratch around, although there may be certain times during the year – such as in winter; when the birds are young; and at the moult – when supplements may prove beneficial.

It is particularly important that sufficient vitamins are included in the diet of your breeding stock, otherwise the resulting chicks may hatch with curly, mis-shapen toes, splayed legs or twisted necks and breastbones. (It is important to note here that similar defects in incubator-hatched birds may, however, be a result of bad incubator management and have nothing to do with genetic deficiencies.)

When showing birds or moving them long distances, there is always a possibility that they might become affected by stress. This is another situation where supplementation is useful, as a course of soluble vitamins, obtained from your veterinary surgery and given in the drinking water for five to seven days, can help in alleviating stress. As sachets of soluble vitamins have a long shelf life, they can usefully be kept on hand. Be sure to follow the vet's instructions to the letter and discard any medicated water that is more than 24 hours old.

WHEN AND HOW MUCH TO FEED

The amount of food required depends on the breed of your birds, their stage of development, the time of year and the methods and environment you have chosen to house them. Ideally, chickens prefer to feed shortly after descending from the roost and before laying their eggs. After some dusting and exploring, they will then quite often take a rest in the middle of the day, before feeding again in the mid- to late afternoon. If your feeding routine can reflect this, so much the better.

There are generally two options when it comes to feeding: ad lib and by hand. It is important not to overfeed your stock, as any surplus will encourage fat, unhealthy birds that will neither breed nor lay to the best of their ability, as well as the very real possibility of a plague of vermin. Most breeders prefer a programme of hand-feeding, as the correct quantities can be judged more accurately. However, where this is not possible (perhaps owing to work commitments), you may have

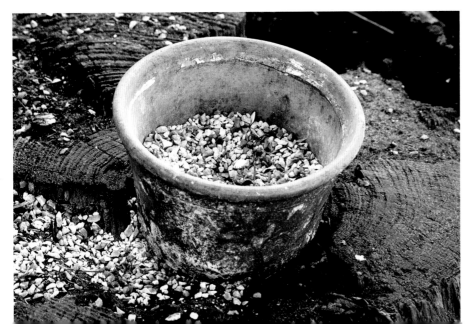

It may not look very appetizing, but mixed grit is essential to the chicken's digestive system.

to compromise, with pellets or mash fed in a hopper in the morning and an evening scattering of grain. Remember that chickens are daytime-loving creatures, so the last feed should not be given so late that they have insufficient time to fill their crop for the long night ahead before going in to roost.

When estimating how much to feed by hand, work on the basis that a fully grown hen eats about 100g (4oz) of food a day, of which half (a good handful) should be grain. Particularly large breeds may require a little more – say, 150g (6oz). For adult bantams, a 60:40 ratio of pellets to grain is considered by many breeders to be a better mix, but it is up to you to observe their daily food intake carefully in order to determine the correct quantity. If you find no trace of food an hour or so after the birds have been given their breakfast, increase the amount until the last particles are being picked up around lunchtime. Conversely, if there is food left from the morning when you take out the afternoon grain supplement, decrease the breakfast quantity accordingly. Very young chicks should be fed four or five times a day, but once they are reasonably active and strong, a small trough or hopper can be used for ad-lib feeding.

Right: Plastic containers are less durable than galvanized ones, but easier to clean.

Far right: Drinkers are best placed outside the run so that floor litter does not become sodden or get scratched into the water.

FOOD AND WATER CONTAINERS

When considering the size of feeder required, the important measurement is the outside edge: for a circular feeder, allow 2.5cm (1in) of the perimeter for each bird; and for a trough, increase this length to 10cm (4in) per bird. For reasons of biosecurity (see page 144) it is best to feed your birds indoors, but if you need to feed them outside from a hopper, do not be tempted to buy one with a flat feeder pan. Feeder-pan bases must be conical so that the food flows into the trough freely and no unused or mouldy feed remains in the centre.

If you are using troughs in the run or henhouse, these should be mounted on bricks or wall brackets so that they do not become contaminated with floor litter. For the same reason, circular feeders should be suspended from a cord or chain attached to a hook fixed to one of the roof beams.

Drinkers may be galvanized or plastic and come in several varieties. Allow 2.5cm (1in) of drinker perimeter per bird. Fountain drinkers are more hygienic on a day-to-day basis, but drinking troughs can be advantageous in the winter as the water they contain can be prevented from freezing over by floating glycerine on the surface each morning when you change it. For very small numbers of chickens or for bantams, earthenware dog bowls make good feeders and drinkers, as they are heavy enough to avoid overturning and are easy to keep clean.

Another option is to construct an automatic drinking system, which can be done quite easily by fitting a ballcock and valve into a heavy-duty plastic bowl. Water is supplied via an alkathene pipe fed from a mains water supply. It is important to check that the ballcock is working freely in order to prevent flooding, and to place the

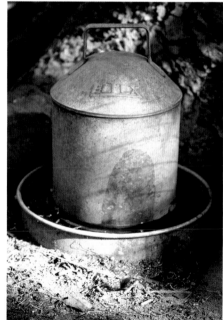

bowl on a level surface. Alternatively, low-pressure hanging drinkers can be obtained inexpensively and easily from any agricultural stockist. Suspend them from a tripod made of stout stakes or hung from a crossbar attached between two posts. Water is supplied from a header tank, not directly from the mains, and the filter that prevents sludge from fouling the spring-loaded release washer must be cleaned weekly. Both types of automatic drinker can be kept clean simply by turning them over and tipping the contents into a bucket so that the ground does not become soggy; they will then automatically refill and be flushed through.

FOOD STORAGE

The shed where you keep your chicken feed should be cool, dry and well ventilated, and the feed must also be protected from dirt, dust and vermin. Provided that family pets cannot gain access to the shed, it is a good idea to keep several baiting points supplied with rat and mouse poison.

Do not store feed bags directly on a stone or concrete floor, no matter how dry it appears, as condensation and moisture will undoubtedly build up between the floor and bag. This is the reason why agricultural merchants store their food on pallets or shelving.

Do not use any manufactured rations that have gone past their expiry date. Not only is there a danger that such food will have gone off; any beneficial drugs or vitamins it contains will also be far less effective or even totally ineffective. If, for whatever reason, any food becomes damp or spoilt, discard it immediately.

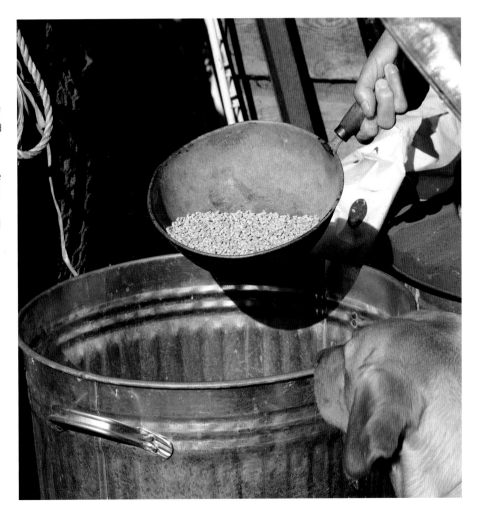

If only relatively small quantities of food need to be stored, keep them in containers that come complete with lids – galvanized ones are preferable, as rats can chew through even the most substantial plastic. Second-hand metal food-storage bins can often be picked up at local farm sales and auctions. Some of these even have dividing partitions, enabling you to keep pellets and grain in the same container.

Dustbins make inexpensive, vermin-proof containers for chicken feed.

BREEDING

Buoyed by your enthusiam for keeping poultry, there is sure to come a time when you will want to rear a few chicks of your own. Whether you want to replace aging birds, or just have some fun, the most rewarding option is to breed chickens from your existing stock. If you don't have a cockerel you could try borrowing a cock bird of proven pedigree and running it with your flock of hens for a month before returning it to its owner. Don't forget to explain your plans to your neighbours, as they might not be too keen on the morning crowing! A quieter alternative would be to buy fertile eggs from a breeder and hatch them under your broody hens.

BREEDING PENS

Assuming that you own or can borrow a cock bird, he should be of the type described in Selecting a Breeding Cock on page 17. The hens must be mature and should conform to the breed standard if there is one. Body size is important, and it should go without saying that all prospective breeding stock must be healthy.

The cock-to-hen ratio will depend on the breed; some males are virtually monogamous, while others are capable of fertilizing the eggs of ten or more hens (see the Choosing a Breed chapter beginning on page 18 for more about this). On average, however, one cock to six hens is about right, although you should seek the advice of an experienced breeder who has a good knowledge of your particular breed. If you want to produce birds for the table or for eggs, flock mating can be quite successful and involves running a number of males with selected groups of females at the rate of one cock to 10–15 hens, depending on the breed. Normally, however, single-male mating (not to be confused with single mating; see box on page 135) is practised.

If all your birds have been bought from the same breeder, there is a good chance that they will be related. As long as they are not too closely related, there should be no problem with this, but repeated inbreeding can result in undesirable traits in your strain (see box on page 135).

Cock birds can be run with the flock year-round, although professional breeders sometimes house them separately until the breeding season, believing that this keeps the males keen and eager to do their job. If the cock is kept separate, introduce him to the hens at least a month before fertile eggs are needed for hatching. Eggs may be fertilized within a week of mating, but it is best to allow a little more time to be sure. If a male of another bloodline is subsequently introduced, it is advisable to allow a fortnight after the changeover before beginning to save any eggs for hatching.

THE FERTILE EGG

If you have bought in fertile eggs, leave them for 24 hours so that the yolks can settle before incubation. They do not need to be kept warm at this stage, as development will simply cease if the surrounding temperature is below 20°C (68°F). If the egg is then brought back up to a suitable temperature within a reasonable time period, either under a broody hen or in an incubator, the arrested development will resume and proceed as normal.

The length of time a fertile egg remains viable depends on the conditions in which it is stored. To obtain the best chances of hatchability, collect eggs as quickly as possible and store immediately at a cool temperature, as they will keep longer than those cooled more slowly through being left in the nest boxes. The ideal storage temperature is 12.7°C (55°F). The storage time also depends on whether the egg is going to be hatched artificially or naturally with a broody hen. Generally speaking, eggs that will be hatched under a broody can be kept for around ten days, whereas those going into an incubator should not be more than a week old. It is interesting to note that chicks from eggs that have been stored are usually smaller than those from eggs that have not, and they also tend to hatch several hours later.

BROODY HENS

You might notice that one of your hens stays in the nest box longer than usual. If she remains there for a couple of days, put your hand under her; if she tries to peck at it, ruffles her feathers and squats further down into the nest, she is broody and is ready to hatch some eggs. However, before you give her fertile eggs to hatch, check that she is of a suitable type: most of the dual-purpose fowls sit well, as do hens of cross-bred stock from heavy breeds.

The best location for a sitting box is in a shed well away from other birds. This should be well ventilated, fairly dark and free from rats – an old-fashioned coop and run is ideal. Cut a grass turf the same size as the box and place it upside-down in the bottom. This prevents eggs from rolling out when the rest of the nest is built up with hay, and it also helps to retain moisture and humidity, both of which are important for incubation. Do not move a broody hen from the henhouse to the sitting box in daylight; instead, wait until dusk.

A good 'broody' will keep a clutch of chicks warm and safely tucked under her feathers.

Only when a hen is sitting tight in her chosen nest should she be considered as a broody and given a clutch of eggs to hatch. A hanging basket is not, however, the best example of where to hatch a clutch of eggs!

INCUBATORS

Hatching eggs in an incubator can be quite complicated and the results are nowhere near as successful as those you will achieve using a broody hen. There are, however, many small electric incubators on the market that may be worth experimenting with. If you are intending to rear poultry as a commercial sideline at some future date, an incubator will be essential – you cannot guarantee that broody hens will be around when you want them. Each incubator model is different, so it is important that you read and understand the manufacturer's instructions if you are to have any hope of success. If you want to hatch just a few eggs, a member of your local poultry club may be prepared to incubate them for you alongside his or her own eggs.

Although there are many incubator models, there are only two basic types: cabinet machines and those that work on a still-air principle. The latter are used widely and very successfully, have few working parts that can go wrong, are easy to understand, and are inexpensive and simple to maintain. Most of the modern models are electric, although you may still come across an older oil- or paraffin-heated version for sale at a farm auction. Ask someone experienced to check it over for you and make sure that all the essential working parts are present before you make your purchase.

Cabinet machines usually have more than one level of setting trays and differ from still-air models in that a fan circulates the

Make sure she is free of mites and lice (see pages 150–1), giving both her and the nest a good dusting with flea powder before she settles down.

Rather than risking fertile eggs straight off, it is a good idea to put a dummy egg in the nest and then give the hen a day to acclimatize to her new surroundings. If, after 24 hours, she is sitting happily and tightly, encourage her out to feed and drink. You can then replace the dummy egg with a clutch of real ones. Just how many eggs to incubate will depend on the size of the hen, but it is often thought that an odd number should be set, as this makes it easier for her to turn them.

The broody should leave the nest once a day to feed and to empty her bowels. She may be so intent on carrying out her matronly duties that you have to lift her off – if so, make sure you feel around her when you pick her up in case any eggs are tucked in her feathers. Keep an eye on her and do not let her stay away from the nest for more than 20 minutes.

air constantly to ensure an even temperature. Because of their size and capacity, they are normally used only by commercial chick producers. Cabinet machines also require a separate hatcher to which the eggs are transferred a couple of days before they are due to emerge from the shell.

Humidity is an important aspect of incubator management, as an egg loses water through evaporation from the moment it is laid. During incubation, this rate of evaporation must be controlled by adding water according to the manufacturer's instructions. Too great a loss of water can kill the embryo, while too much can drown it. Ventilation in the incubator is also crucial and serves two purposes: supplying oxygen and removing harmful gases such as carbon dioxide. Some machines have several felts, which are removed one by one at intervals of a week and let out the stale air.

Most modern incubators have fully automated turning devices, although you must check daily that the turner is working by noting the position of the trays or eggs. In older models, the eggs must be turned by hand at least twice a day to prevent the developing embryo from sticking to the side of the shell.

Generally, the calm heavier breeds make the best broodies and mothers.

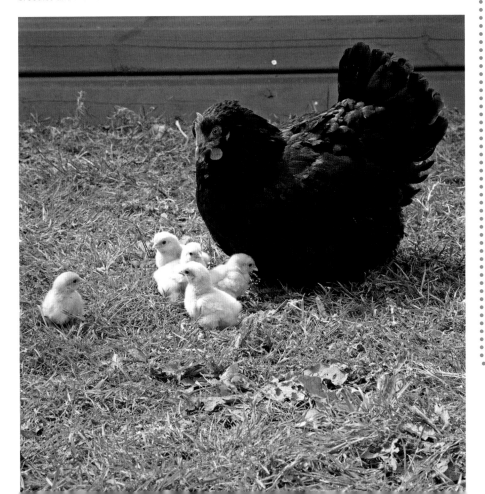

COMMON BREEDING METHODS

Line breeding uses two unrelated birds, the offspring of which are subsequently mated together in a systematic way to produce two distinct lines of chicken based on each of the parents. The lineage of any offspring, no matter what generation, are traceable to one or other of the original parent birds and, in theory, the gene pool of the first generation can be re-created endlessly.

Inbreeding is used to fix desirable genes, although it is not recommended for the novice as undesirable genes may also be passed on. The system is similar to line breeding, but brother–sister mating is used.

Outcrossing is when a totally unrelated bird is brought in to add new blood to an existing strain in situations where that strain has become weak through repeated line breeding or inbreeding. Like inbreeding, outcrossing can be risky, as unwanted traits may be introduced if the new bird has not been selected with the utmost care.

Single mating involves mating a single male to a single female and is uncommon except in the practically monogamous Asiatic breeds such as the Malay. It may also be the first step in a line-breeding strategy or as an outcross. (Double mating has a totally different meaning and is explained on page 17.)

CANDLING

When an incubator is being used, it is advisable to check the eggs each week to ensure that the embryos within them are developing correctly. This is best achieved by candling, which involves holding each egg below eye level in front of a bright light, preferably in a darkened room.

Candling lamps can be bought or made inexpensively and consist of an electric light bulb placed inside a box, on top of which is an egg-shaped hole. The egg is held over the hole, such that no light escapes around its sides (most candlers have a rubber seal).

It is recommended that candling is carried out weekly to eliminate infertile, dead or damaged eggs. It is also a useful method of checking humidity by monitoring the air sac. At the first candling, at five to seven days, a fertile egg is indicated by the presence of the blood vessels of the embryo (these resemble a multi-legged spider), and the air sac should be evident at the broad end of the egg. As incubation progresses, the air sac becomes larger until, by hatching time, it takes up nearly a quarter of the egg, the remainder of which will appear almost completely blacked out. If the air sac is deemed to be too large at any stage, the humidity level inside the incubator must be raised. If the air sac is too small, you must lower the humidity. This can be done either by increasing ventilation or by reducing the surface area of any water trays in the incubator.

At 14 days, infertile eggs will appear clear and should be removed from the incubator to prevent possible contamination. This also improves the chances of the remainder hatching.

Candling and the whole process of incubation is too complicated to discuss in depth in a book of this nature, and it is therefore recommended that anyone interested in the subject reads the comprehensive, revealing and informative *New Incubation Book* by Dr A.F. Anderson Brown and G.E.S. Robbins (Blaine, WA: Hancock House Publishers, 2002).

A day or two after hatching, chicks are becoming curious about the outside world, but they will need to scuttle back to 'mum' for about three weeks in order to keep warm.

RAISING CHICKS

Once the eggs begin to hatch, confine the broody hen to the nest box and leave well alone. As soon as the hatch has finished and the chicks are dry, transfer them to a coop along with the hen. If the coop is in a shed, make sure that the hen cannot take her chicks too far from the nest by building a temporary pen. The sides of this must be tall enough to prevent her from flying over, otherwise the chicks will be left behind and may become chilled. If the chicks have been hatched in an outside coop, it should now have a run attached to it and be moved on to fresh, short grass. Feed the chicks on proprietary chick crumbs for the first three weeks and make sure their water container is shallow enough that they cannot drown.

If the chicks have been hatched in an incubator, it is possible to persuade a broody hen to become a foster mother to them. The hen must have been sitting on a clutch of dummy or even infertile eggs for at least a week and be well settled into the nest. Plan the arrival of the chicks for late evening and make sure that they can be kept warm for a couple of hours – buy a chick box for the purpose, or make one by placing a cloth-covered hot-water bottle in the bottom of a cardboard box. As dusk approaches, very gently take away a couple of the eggs under the hen and replace them with two or three chicks. The hen will not like being disturbed, and she will fluff up her feathers and may even try to peck the back of your hand as you slide the chicks underneath her. Tuck them in among the remaining eggs, back off slowly and leave her to settle again. After an hour or so, if all is well, take away the remaining eggs and replace them with the other chicks. For maximum success the process must be carried out slowly and quietly.

Opposite: Candling is shown here in a commercial production facility, but the basic principles are the same for the small-scale poultry breeder.

USING AN ARTIFICIAL BROODER

If you do not have a broody hen to rear your incubator-hatched chicks, you will need an artificial brooder. This is usually located in the sectioned-off corner of a shed, with warmth provided by gas heaters or infrared lamps. If you have only half a dozen or so chicks, a conventional light bulb fixed in an upside-down terracotta plant pot will give off sufficient heat and the chicks will cluster happily around it.

Whatever heating system you choose, the room, shed or building must be maintained at a constant temperature, so forward planning is required. A metal roof, a sunny day and the heat from a lamp, for example, will obviously be a bad combination. Even with suitable surroundings, always keep a check on the ambient temperature inside the brooder, especially around midday.

Chicks need to learn where the source of heat is straight away. If rearing is carried out in a big shed, this is best achieved by making a circular surround of cardboard or hardboard and keeping the chicks within this. Suspend the heat lamp from the ceiling at a height above the floor that ensures the temperature there is around 32°C (90°F). Increase the lamp's height progressively each week until, by the third week, the temperature at floor level is down to 21°C (70°F). You can usually tell if you have got the temperature right by watching the chicks closely: if they are huddled under the heater, they are too cold; if they are seen pressing against the surround, they are too hot and the ring should either be enlarged or the heat lamp raised. After a few days, increase the cardboard surround to give the birds more space, until

you eventually remove it altogether.

Once the chicks have reached the age of three weeks, the heat source can be turned off in the day during spells of warm weather. If only small numbers of chicks are being reared and the weather is hot, you could take the birds to a small wire pen on the grass outside for a few hours. However, make sure that the pen is cat-proof, situate it in the shade and take the chicks inside immediately should it start to rain.

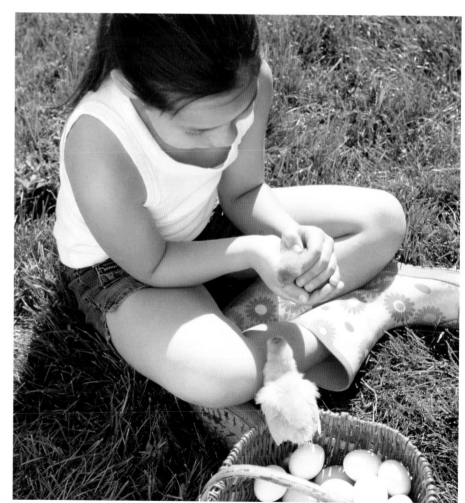

Healthy eggs are more likely to produce healthy chicks.

FEEDING ARTIFICIALLY REARED CHICKS

Feeding and drinking vessels in artificial brooders should initially be placed around the heating area so that the chicks do not have to go far to look for them. Small jam jar-type drinkers are ideal, although commercial plastic chick drinkers are available. Chick feeders can also be bought, but almost any shallow pan will do provided that the birds aren't able to scratch their food all over the place.

Do not worry if you don't see the chicks eating during the first 24 hours, as they still have enough nutritional reserves from the egg yolk. However, as they don't have a mother to show them what to do, you might have to encourage them to eat yourself by 'pecking' at the food with your finger. Some finely chopped hard-boiled egg sprinkled over the chick crumbs will help to arouse their curiosity and also provides extra protein for the first couple of days. Do not mix up too much at any one time – as with wet mash for older birds, it will soon go sour.

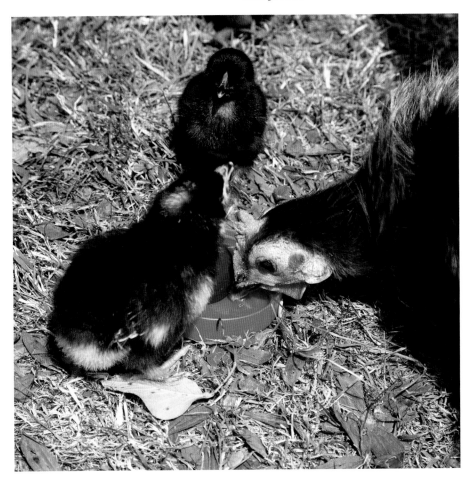

These chicks have a mother to show them what to do. If you don't have a suitable hen, you may have to give 'pecking lessons' yourself.

EGGS

For centuries the riddle of the chicken and the egg has led philosophers to examine how life and the universe began. Which came first? Was it the hen, which after all lays the egg, or was it the egg, from which the hen hatches?

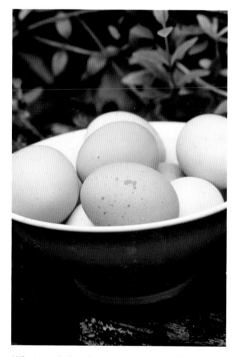

Whatever their colour eggs are one of the most nutritious foods money can buy.

IN THE BEGINNING...

The Bible and the theory of evolution provide different answers to this question, but whichever there is plenty to wonder at in the remarkable, wholesome, miraculous egg.

Eggs provide all the essential amino acids and minerals required by the human body, and are an important source of vitamins A, B and D. They also supply a complete, high-quality protein, while eggs from hens fed a flaxseed-rich diet provide omega-3, which helps to lower blood triglyceride levels, promotes good vision and is required for normal growth and development in the human body.

Some people avoid eating eggs in the belief that the high cholesterol levels found in the yolk are harmful. It is thought, however, that egg yolk may actually lower levels of low-density lipoprotein, the undesirable form of cholesterol, while raising levels of high-density lipoprotein, or 'good cholesterol'.

EGG FORMATION

An average egg weighs 60g (2oz) and consists of three parts: the shell, which is made of calcite (a crystalline form of calcium carbonate); the albumen, or white, which is made up of water and protein; and the yolk, which is the most nutritionally valuable part of the egg.

Eggs develop in the ovary individually, then detach and slip into the oviduct, a long tube that ends in the vent, or cloaca, through which they exit the body (droppings are passed through the same orifice).

The hen's oviduct has two parts: in the first, the white forms around the yolk; and in the second, the shell is made and the pigment of the shell is deposited. Each egg takes 3–5 hours to pass through the first part of the oviduct and a total of 15–20 hours before it is ready to be laid.

While the colour of the shell has no relevance to the nutritional value of the egg, the colour of the yolk is directly associated with the hen's diet. A very pale yolk can indicate that she lives in overcrowded quarters, is underfed or lacks greenstuff, whereas a bird fed a diet rich in xanthophylls (the yellow pigment from the carotenoid group found in green leaves) will produce a darker yellow yolk.

Occasionally, a hen may lay an egg with no yolk at all or one with a double yolk. These are both the result of unsynchronized production cycles, and usually occur at the start or end of a laying period. The so-called 'meat spot' in the egg is, in fact, a small deposit of blood caused by the rupture of a blood vessel during its formation. It does not indicate a fertile egg. It is not harmful, but can easily be removed with the tip of a knife.

A cloudy white is a sign of freshness and is caused by carbon dioxide – this gas is present naturally in the white when the egg is laid, and decreases over time as it escapes through the porous shell.

EMBRYO DEVELOPMENT

All birds will lay a clutch of eggs before starting to incubate them, and a hen is no different in this respect – she will just continue

EGG GAMES

Egg jarping, or egg tapping, is a traditional Easter game from the north of England and is rather like conkers. Players tap each other's uncooked eggs in turn until one breaks. The victor then goes on to the next round, until eventually there is only one good egg left – the winner.

•

Egg rolling takes place on Easter Monday, and again is particularly popular in the north of England. Hard-boiled eggs are rolled down a hill, and the winner is the person whose egg rolls the furthest, survives the most rolls or lands nearest a target, depending on the variation played. Easter egg rolls are also popular in the US – the event on the lawn of the White House is the best known and has become an annual tradition. The first official White House Easter Egg Roll was held in 1878; today, the President and First Lady invite children from across the country to attend the celebration.

to lay as long as her eggs are removed daily and she thinks she hasn't produced enough for a clutch. The clutch must be complete before incubation starts, so each day the hen will add an egg and then leave it to cool. When enough eggs have been laid, she will go broody and will lose her breast feathers in order to warm the eggs and begin incubation. Her behaviour also changes: she will remain on the eggs, fluffing up her feathers and making croaking noises if approached, and will leave the nest only once a day to eat, drink and defecate.

The embryo of a large fowl normally develops inside the egg for 21 days, whereas that of a bantam has a slightly shorter incubation period (19–21 days). As it grows, its primary food source is the yolk. It is important that the embryo doesn't become

stuck on one side of the egg, so the broody hen will turn all her eggs several times a day. After 21 days, the chick pecks its way out of the shell with a horny growth on its beak known as the egg tooth.

Once the chicks have hatched, the broody hen will remain on the nest for a further 24–48 hours. After this time, any eggs that have not hatched will be left behind when she takes her chicks on their first outing.

Certain breeds of chicken – in particular the Mediterranean varieties such as Leghorns and Minorcas – have had their broodiness bred out of them so that they produce more eggs. It is also rare for hybrids to go broody. As you may not always want to be hatching chicks, there are various methods of dissuading a hen from being broody. Sometimes, just isolating her away from the nest box will do the trick. A more stubborn hen can be put in a cage with a wire bottom, so that the airflow cools her underside and discourages her from sitting, or you could try giving her a 'clutch' of ice cubes – although it may take several attempts before this cure proves effective.

EGG SIZES

Egg sizes in the UK now follow the European convention:

Small (S): less than 53g (1.9oz)
Medium (M): 53–63g (1.9–2.2oz)
Large (L): 63–73g (2.2–2.5oz)
Very large (XL): greater than 73g (2.5oz)

EGG STORAGE

A fresh egg will keep quite safely for five to six weeks in a cool, dark, dry place. Always store eggs with the point down so that the yolk stays centred. As the porous shell will absorb smells from its surroundings, eggs should also be kept away from strong-smelling foods such as fish or scented products like soap. The porosity also makes it unwise to wash eggs – if they are very dirty, brush or wipe them instead.

Many people keep their eggs in the refrigerator, but in fact they are better off in a cool place but not chilled. Refrigerated eggs frequently break when boiled, the absorption of bacteria through the porous shell is increased in the enclosed space, and the lower temperature can cause the protein to break down, reducing nutritional benefits. Take your cue from supermarkets and other food shops, which always store eggs on unrefrigerated shelves.

Raw eggs can be frozen very successfully and kept for up to a year. If you freeze them whole, the shells will crack, but the eggs can still be used for baking if defrosted carefully. A more successful way of freezing eggs is to put spare whites or yolks into ice-cube trays, defrosting the quantity you need as necessary. Cooked eggs do not freeze well, as they tend to be tough and rubbery when defrosted.

To tell if an uncracked egg is fresh, drop it into a bowl of water. If it sinks, it is fresh, but if it floats it is bad – bacteria have entered through the shell and have created gas inside. To tell if an uncracked egg is raw or hard-boiled, spin it, then abruptly stop it from spinning and immediately let go. If it starts to spin again, it is raw as the liquid inside will continue to rotate. If it remains still, it is hard-boiled.

RULES AND REGULATIONS

If you intend to keep large numbers of birds or plan to sell your eggs anywhere other than from your home, there are a number of regulations you must abide by. Some of these rules relate to food safety and egg traceability, while others are concerned with bird health.

In the UK, if you own more than 350 hens you must be registered with the Egg Marketing Inspectorate (this is free) and your eggs must bear a date stamp and your producer registration number to allow traceability. The stamp must be legible and applied with food-grade ink. The eggs must also be graded by size, which can be done only at a registered packing station. If you own fewer than 350 hens but wish to sell your eggs, they must also be stamped with a producer registration number and you can only sell them ungraded. You do not have to mark eggs sold at your farm gate or delivered by hand to your customers, but they cannot be used in the catering industry.

Producer registration numbers are used throughout the European Union and look like this: 1UK 56789. The first number refers to the way the hens are reared: 1 = free-range; 2 = barn-reared; and 3 = caged. The letters refer to the country of origin, while the final five digits are unique to the producer.

If you keep more than 50 birds of any breed or species commercially in the UK, you must register them on the Great Britain Poultry Register. This is part of the programme set up by the Department for Environment, Food and Rural Affairs (Defra) to improve risk assessment and monitoring of

EGG-EATING

Egg-eating is a habit often acquired through simple curiosity. It starts when a thin- or soft-shelled egg is laid or an egg is broken and one of the birds in the flock pecks at it and decides it likes the contents. It will then do the same to any other eggs it may subsequently find.

•

Once egg-eating has become a habit, it can be very difficult to stop. Clean, dark nest boxes will keep eggs out of sight and therefore reduce temptation. In addition, blown eggs filled with strong mustard can be put in the nest boxes and may taste unpleasant enough to cure a bird, although you may have to do this for several days before the message gets through.

bird flu outbreaks. Bird flu poses no danger to egg consumers, as the virus cannot survive cooking, and often one of the first signs of the disease in a hen is that she stops laying.

To entitle you to call your eggs organic, the layers must have access to an outside area all year round or be fed sprouted grains for any period they are kept indoors. All feed must be certified organic and cannot include meat by-products, and no antibiotics can be given. In the UK, eggs denoted as class A are the highest quality possible and may not be cracked, should have a normal shell and should not be washed – it is considered preferable to produce a clean, quality egg in the first place, as this indicates high standards of production. Class B eggs are of a lower quality and may be cracked or dirty.

EGGS IN FOLKLORE

There are many superstitions surrounding eggs. It was believed that to find a small egg was bad luck. Even unluckier was an egg with no yolk, as this was said to have been laid by a cockerel. Sailors were advised never to mention the word 'egg' if they were to avoid misfortune at sea. And anyone eating a boiled egg should poke a hole through the bottom of the shell when they have finished to prevent a witch from using it as a boat, but shouldn't throw the shell on a fire or the hen will never lay again and a storm may brew at sea.

If a girl wants to find out who she will marry, she should boil an egg, fast for a day, then take out the hard-boiled yolk and fill the hollow in the white with salt. While reciting an incantation to St Agnes, she must eat the salty egg, including the shell. The next man she sees will be the one she will marry.

While commercially available eggs are almost invariably brown, keeping your own chickens can give you the pleasure of eggs in beautiful shades of chalky white, blue-green or olive.

EGGS AROUND THE WORLD

Thousand-year-old eggs are a highly prized Chinese delicacy, but despite their name they are, in fact, only around ten weeks old. They are preserved in a coating of clay, ash, lime and salt, during which time the chemicals in the clay soak through the shell, turning the egg a translucent blue or green colour and producing a slightly fishy taste.

In imperial Russia, giving eggs was not restricted to Easter. The practice was so popular that the Romanov royal family employed jeweller Carl Fabergé to create wonderfully ornate gem-encrusted eggs, to be given as presents at any celebration.

The word 'cockney' is thought to have derived from 'cock's egg', the name for the small malformed egg occasionally laid by a young hen. The term was applied by country folk to townsfolk generally, because of their reputed ignorance of country life and customs (cocks don't lay eggs), and in the 17th century specifically came to denote someone born within the sound of Bow bells – the bells of the Church of St Mary le Bow in the City of London.

HEALTH & CARE

Good hygiene is essential to keeping your birds happy, healthy and productive. This is easily achieved by keeping accommodation clean, giving the birds plenty of room and feeding them a balanced diet. Maintaining biosecurity is also an important factor because, as clean as your chicken yard appears, disease can still be brought in on the wind, via the droppings of wild birds, or on the feet of human visitors. Disinfectant is therefore a simple but vital component to keeping your chickens safe.

WARNING SIGNS

The best way to check the health of your birds is simply to sit and watch them for a time each day. Most chickens like to be part of the flock, so an individual moping quietly in a corner on its own is usually an indicator that all is not well. The birds should always be busy scratching, dusting and feeding. Their combs should, in most breeds, be red and waxy and their eyes must always be bright. Drooping wings, ruffled feathers, a loss (or sudden gain) of appetite and loose droppings stuck to the feathers around the vent area should all be treated with great suspicion.

If the birds are tame enough, occasionally pick them up to check under their wings and in other downy areas for signs of fleas, lice and mites. At the same time, feel each bird's breast for any sudden loss of weight.

BIOSECURITY

Use of the term 'biosecurity' has become more common since concerns over avian influenza (see page 150) have come to the fore. It is, however, really just a buzz-word for common-sense health measures. By periodically removing droppings from the run, feeding chickens in an area that is not accessible to wild birds and placing a disinfectant foot-bath at the gate to your poultry yard, you are actually carrying out most aspects of biosecurity recommended by government experts.

HANDLING CHICKENS

You will need to handle your chickens from time to time, to move them or to inspect them at closer quarters. First, however, you need to catch them. You may be able to do this by putting down some food and then quickly grabbing them by the legs, or you may be able to drive one bird into a corner and catch it that way. If all else fails, you will be able to catch them once they have gone in to roost for the night.

Birds should never be handled roughly, as this will cause stress and may even damage them physically. To pick up a bird, place both hands over its wings, then lift and place it under your arm so that it is facing backwards. Keep hold of both legs with your hand and keep the wings firmly closed between your elbow and your body.

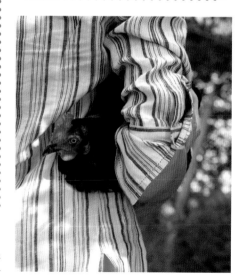

Hold a chicken firmly but not roughly, so that it feels secure.

BIRD CARE

Birds for exhibition will require some extra care in the days before a show to ensure that they are looking their best.

Washing Birds

Only soft-feathered birds that are to be exhibited are normally washed; washing hard-feathered varieties such as the Old English Game spoils the tightness of the feather conformation, so traditionalists 'polish' them with a silk handkerchief instead. Washing can also occasionally be necessary in breeds that have crests or feathered feet.

For showing purposes, wash the bird several days before the event so that the feathers have time to dry fully and hang properly. Most fanciers use two bowls of lukewarm water, the first ideally containing pure soapflakes, although washing-up liquid or baby or dog shampoo will do almost as well. Whatever soap is chosen, use only a little or it will strip the natural oils from the feathers. The second bowl should contain clean water for rinsing, with a little clothes whitener added for white or light-coloured breeds.

First, wash the feet and legs separately outside the main bowl, so that the dirt does not go into the clean water. A firm toothbrush or nailbrush is ideal for this, but any dirt that is ingrained under the scales can be picked out using a cocktail stick or a toothpick.

Next, holding the bird firmly so that it feels secure, immerse it in the first bowl of water and, starting at the head, very gently work the lather through the feathers in their direction of growth, taking care not to get soap in the bird's eyes. Once they have got over the initial panic of being placed in the water, most chickens seem to like this part and will crouch quietly in the bowl while they are being washed.

Rinse the bird clean in the second bowl of water, then pat it dry with kitchen roll or old towels; some exhibitors use a hairdryer, although if held too closely this can damage the feathers. The bird can then be left to dry completely in a box or poultry basket in a warm room or near a radiator. Once it is thoroughly dry, a show bird should then be put in a show pen in the penning room. Non-show birds can be returned to the flock, although remember that the washing process will have removed some of the feathers' natural oils and so the bird may have less protection against inclement weather for a day or two.

Showing Tips

Every experienced poultry exhibitor has his or her own methods of enhancing a bird's appearance. For some this involves shaping and drying the feathers in a particular way after washing, while others may use rouge or baby oil to brighten a red comb, baby powder to whiten ear lobes or petroleum jelly to smarten up legs. There are a multitude of other tips that you will find out only by befriending a successful breeder, although he or she is unlikely to want to divulge any ingenious tricks of the trade if you are going to be in direct competition.

Finally, after a show, birds should always be quarantined for a few days before being returned to the flock to prevent the possible transmission of disease.

Anyone who is interested in showing chickens should read David Scrivener's comprehensive and informative book *Exhibition Poultry Keeping* (Crowood Press, 2005).

Healthy birds in a hygienic environment will keep themselves clean through dust bathing and preening.

Moulting

Most birds shed their feathers annually in the late summer or early autumn, although young birds moult twice during their first six months of life. A partial moult sometimes also occurs in the early part of the year. This is often limited to the neck, especially in the case of point-of-lay pullets that have been given layers' rations too early on in their life.

A healthy young chicken will take around six weeks to change its feathers; older birds can take double this time. Some of the lighter breeds take less time than heavier, dual-purpose breeds. The large wing and tail feathers of all birds are replaced slowly in a specific sequence that has evolved over thousands of years to ensure that birds do not lose too many feathers at once and so are always able to fly and escape from danger.

It is better for the bird if the moult takes place quickly, and in order for this to happen it must be in good condition. The later in the season a moult starts, the longer the whole process will take. And because egg-laying decreases or even ceases altogether in moulting birds, a laying bird will remain unproductive for longer.

In order to get egg-laying breeds back into production as soon as possible, an early moult can be artificially induced by moving the birds into a different environment or by radically altering their diet. As soon as most of the hens in the flock are moulting, restore their usual feeding routine and give them an extra boost of linseed meal, soybean meal and cod-liver oil. Once they have produced their new feathers, increase the animal protein in their diet to encourage laying.

KILLING CHICKENS

There will come a time when, because of illness, old age or taking a bird for the pot, it will be necessary to kill one of your flock. This is one of the least attractive aspects of the hobby and it may seem as if you are acting as jury, judge and hangman when making the decision.

Rather than unsettling the bird in question by chasing it around the henhouse or run during daylight, it is probably best if you can take it off the roost at dusk then remove it from the vicinity of its companions immediately. To ascertain whether or not a chicken is ready for the pot, feel for the two small pieces of cartilage on either side of the breastbone – these will tighten up with accumulated fat when a fowl has been properly fattened. Blowing or parting the breast feathers makes the layers of fat easy to see, and the back should also feel fleshy and not bony to the touch.

The most widely used and arguably most humane way of killing a bird is to wring its neck. For a right-handed person, this is most easily done by holding the bird's feet in the left hand so that it is hanging head down with the breast towards you and the back facing away. Next, grab the head with your right hand, so that the neck is between either your two middle fingers or your thumb and first finger, and the head is in the palm of your hand. With a sharp movement, pull the head downwards and simultaneously rotate your wrist upwards, so the neck is stretched tight and the head is pulled back. You should feel the neck dislocate and may even hear a light crunch as the vertebrae break. Don't worry if the bird goes into violent spasms and flaps its wings for some seconds afterwards, as this is simply the nerves twitching in death.

PREPARING A BIRD FOR THE TABLE

Birds that are intended for the table should be starved for 24 hours before killing, after which they need to be plucked, stubbed, have their head and feet removed, and be drawn (eviscerated).

Plucking

The plucking should be started while the bird is still warm. Remove the flight and main tail feathers, then, sitting comfortably, hold the bird across your knees and start on the breast. Hold the skin tightly with one hand to prevent tearing, then with the other pull small groups of feathers out against the natural direction of their lie. Proceed in a similar manner over the whole body.

Any stubbly feathers (caused by new growth on a young bird) can be removed once the bird has cooled and the skin has begun to set firm. The bird should then be allowed to cool before the internal organs are removed.

Removing the Head and Feet

Turning the bird on to its breast, slit the skin vertically from the back of the head to its 'shoulder blades' and expose the back of the neck vertebrae. It should be possible to identify the break that occurred at killing by the presence of a small cavity of congealed blood; cut the head off at this point and, feeling with your knife, cut through the neck at its base where it connects to the body. You should then be left with the neck as a separate piece for your stockpot and, on the chicken, a flap of skin folding over from the front.

Break and twist the legs at the rounded joint where the scaly part attaches to the thigh (you may need to cut around the joint with a knife), and pull the feet away from the body. They should come clean away, taking with them the sinews that run through the fleshy part of the thigh.

Drawing

The internal organs are removed by making a small incision between the vent and just below the parson's nose. Cut carefully around the rectum to detach it from the rest of the body. By inserting your hand into the gap created, you should then be able to remove all of the internal organs in one operation. Wipe the insides with a clean tea towel or several sheets of kitchen paper.

If you want to keep the heart, gizzard and liver for stock, separate them from the rest of the organs, which can then be discarded. The gizzard is easily identified by its blue/silver colour and muscled appearance; its insides should be cleaned by cutting three-quarters of the way around its edges and opening it up like a book.

Finally, feel inside the flap of skin that was created by removing the neck and pull out the crop. Also, check that the windpipe (trachea) came away when the neck was detached.

Once prepared in this fashion, the bird is ready to cook, although there is no doubt that its flavour will noticeably improve if it is stored in a cool place for between 24 and 48 hours.

Small birds like this can be held gently but firmly close to the chest.

MINOR AILMENTS AND CONDITIONS

If you do notice anything a little out of the ordinary, it is important not to panic. Your first approach should be to seek the advice of an experienced poultry keeper, preferably by getting him or her to visit, although a simple phone call may provide the reassurance you need. If you are still unsure about the diagnosis or treatment of an ailment, do not hesitate to contact your vet.

Colds

It is possible for chickens to catch colds, although the classic symptom of a runny nose may also be an indicator of other problems. As with humans, the likelihood of infection is increased by exposure to draughts, damp and sudden fluctuations in temperature. Colds will get better on their own, but it is possible to purchase avian cold cures from your vet or agricultural supplier.

Infectious coryza is a bacterial condition that produces symptoms similar to those of the common cold, including laboured breathing and swollen eyelids. Although this ailment is infectious, an antibiotic treatment is very effective.

Crop-Bound Birds

The crop of a chicken can become congested, or crop-bound, for a variety of reasons, the most common being an obstruction caused by eating feathers, litter or long grass. Old-fashioned cures include giving the hen a drink of warm water to distend the crop, which should then be softened by rotating and massaging it by hand. If this does not work, a bird can sometimes be operated on, although in practice it is often necessary to destroy it and open up its crop for analysis. In this way, the cause of the problem can be identified and steps can be taken to prevent it from happening to any other birds in the flock.

Egg-Eating

See box on page 142.

Feather-Pecking

Typically, feather-pecking is the result of overcrowding and boredom. Giving your birds more space, dusting sites and a regular supply of greenstuffs can usually eradicate the habit, although once it takes hold it is sometimes difficult to stop. Very often, just one bird starts feather-pecking as a result of an inquisitive peck, and if she draws blood she and others in the flock will be encouraged to continue. At the very least, the neck, rump and vent will be denuded of feathers, and, at worst, the flock will continue pecking until the victim is so severely injured that she dies. Pecked birds can be sprayed with proprietary brands of anti-pecking remedies and the wounds treated with an aureomycin-based animal powder.

Fleas

Hen fleas are visible to the naked eye. Flea infestation in a bird is sometimes evident from weight loss and the appearance of bare patches in the feathers; the comb and wattles may also appear anaemic owing to blood loss caused by the feasting fleas. Fleas thrive in warm, humid conditions and are therefore more of a summer problem. Treat birds with a flea powder and spray all internal surfaces of the house in order to eradicate the flea eggs, pupae and larvae. A regular reapplication is recommended – as often as every three weeks in hot weather.

Lice

These parasites come in several different varieties, each of which has a preference for a certain area of the bird, but they are most commonly found around the vent area (and in the tufts of crested breeds). By parting the feathers, you may see small, light-coloured lice running between them. Look out for lice eggs in the form of a greyish-white encrustation around the vent and in the feathers under the wings.

Treat affected birds with either a liberal dousing of louse powder or a liquid chemical spray. A weekly dusting of nest boxes will protect laying hens from becoming infested, but any cockerels will have to be treated individually.

Mites

There are several species of mite, the most common being the red mite. Despite its name, this parasite is, in fact, greyish-white in colour, becoming red only when it is full of blood after feeding on birds during the night. Red mites live and breed in crevices in the henhouse as near to their meal as possible – hence their fondness for the ends of perches and the interior of nest boxes. Watch out for the characteristic signs of infestation – 'egg spotting', or blood spots from the squashed mites on the surface of eggs – and inspect the sleeping quarters regularly by torchlight.

Northern mites are similar in size to red mites but are grey or black in colour and live continuously on the bird's body, most often around the vent area. Birds suffering from a really heavy infestation tend also to become scabby around the facial parts. The red mite and northern mite can both be treated with a pyrethrum-based spray, although a single treatment is seldom effective and regular follow-ups are essential.

The scaly-leg mite differs from other mite species in that it affects only the legs and feet. Left untreated, the legs may become swollen and crusty, and eventually the bird will become very lame. The mite burrows under the scales on the legs, creating tunnels where it is able to breed. As it does so, the scales lift and distort. Old-fashioned remedies include sulphur ointment, paraffin and linseed oil. Petroleum jelly also works well, as it seals the gaps in the scales and prevents the mite from breathing, and vets recommend an application of gamma benzene hexachloride.

Prolapse

This condition is rarely seen in intensively reared birds but is quite common in otherwise healthy free-range ones. It occurs in immature or slightly overweight hens that have been laying heavily, and is caused by straining, the result being that the vent muscles are pushed out of the body with the egg still attached inside.

If prolapse occurs, the egg should gently be broken and removed, and the exposed organs then cleaned with a mild antiseptic. After this, they can be coated with a lubricant and pushed carefully back inside the abdominal cavity (this is more easily achieved if the bird is held head down). Keep the hen isolated in a small coop and run for a week, feeding her a diet of grain-based feed. Provided that the problem does not reoccur, she can then be returned to the flock.

Worms

Chickens are affected by two main types of worm: the tapeworm, which is flat and segmented, and the roundworm, which is round and smooth. The life cycle of each parasite usually involves a stage in the environment or in an intermediate host. Most, however, are passed from one bird to another by means of its droppings – fertilized worms are excreted and are then picked up by the other bird. Tapeworm eggs may also pass via the droppings or be retained within the rear segments of the worm, which periodically break off and are excreted. These eggs are then eaten by snails and beetles, which in turn are eaten by the poultry.

Birds suffering from a worm infestation may show an increase in their appetite combined with a decrease in egg production. Combs will look pink rather than red, and birds with a heavy infestation of roundworms will excrete bundles of dead worms. Preventative worming is normally carried out twice a year by the inclusion of a drug in the feed. Flubendazole is the most commonly used vermicide and can be prescribed by a veterinary surgeon, pharmacist or specialist agricultural merchant.

MORE SERIOUS ILLNESSES

Any chickens showing obvious signs of serious ill health must be taken to the vet immediately, and any that die should undergo a post-mortem by a vet who specializes in poultry to determine the cause of death. Unless you are absolutely certain that you have made the correct diagnosis, do not attempt any treatment. While antibiotics are invaluable when used correctly, they should never be given as a standard treatment, as they may merely mask the true symptoms of a disease rather than cure it.

In the UK two diseases affecting chickens – avian influenza (bird flu) and Newcastle disease – are notifiable, which means that if you suspect you must, by law, inform the Department for Environment, Food and Rural Affairs (Defra). It is advisable to keep abreast of any developments concerning these diseases, either through the media or, perhaps more reliably, via the statements periodically issued by interested organizations and the government's chief veterinary officers.

Acute Death Syndrome

The term 'flip-over syndrome' accurately describes this condition of acute heart failure. The causes of heart attack can be many and varied, and as a result they often remain unknown, but overfeeding is undoubtedly a contributing factor – the syndrome is not uncommon in commercially produced table birds prior to slaughter. If, in post-mortem examination, a fatty deposit is noticed around the heart, liver and kidneys, you have a legitimate reason to suspect overfeeding and should alter the diet of the rest of your flock accordingly.

Aspergillosis

This condition results from the inhalation of large numbers of fungal spores, which create lesions in the respiratory tract and cause obvious respiratory distress. Occasionally, an infection of the brain is also noticed in post-mortem examination. There is no treatment, so it is important to clean out damp feed, long cut grass and mouldy hay, straw or wood shavings from the run, all of which form the perfect mediums for the growth of the various fungi involved.

Avian Influenza (Bird Flu)

Bird flu has been given a separate section here owing to growing worldwide concern over the disease. The disease has actually been recognized since 1878, and by 1901 its cause had been identified as a virus and it was named 'fowl plague'. In 1955, its relationship to the mammalian influenza A viruses had been proven, but it was not until the 1970s that it was realized that vast pools of influenza A viruses also exist in the feral bird population. It is this latter fact that makes the control of any serious outbreak so difficult, especially where poultry are kept outdoors.

The avian influenza virus, known as H5N1, breeds in the respiratory and intestinal tracts of infected birds and is transmitted from bird to bird, either by coughing and sneezing or through the faeces. There is currently little evidence of an airborne spread over long distances. The clinical signs are a cessation of egg-laying, a difficulty in breathing or, simply, sudden death.

Routine biosecurity precautions at home to prevent the spread of bird flu include keeping visitors away from your birds, ensuring that all feeding utensils in the henhouse are kept clean in order to reduce visits by wild birds, and adding vanodine or a similar product to the drinking water each day. Spraying the insides of the house on a weekly basis with a virucide will also help.

A vaccine containing an inactivated strain of the closely related H5N2 strain is available. The antibodies produced by birds given this vaccine are also effective against H5N1 because the two strains are so similar. Although birds do need time to build up immunity following vaccination, tests have proven that transmission of the virus is halted after two weeks. The duration of protection is further increased by a booster vaccination given some six to ten weeks after the initial dose. Some countries have opted for a proactive policy of vaccination of domestic poultry, to avoid the need for millions of birds to be destroyed should a serious outbreak of bird flu ever occur. If you suspect that your birds are showing signs of bird flu, you must notify your vet or the appropriate government authority immediately.

Coccidiosis

Normally affecting only young stock, coccidiosis is bleeding within the intestinal walls and is caused by the coccidial protozoan organism. It is spread though droppings, so a good cleaning regime should prevent its occurrence. A natural immunity also tends to develop within a flock, and it has been found that young birds raised on wire floors, where they have no contact with their droppings, are more susceptible. Symptoms include a hunched appearance, ruffled feathers, blood in the droppings and/or sticky white diarrhoea coating the feathers of the vent area. Adult birds may lose weight and condition, bringing about a decrease in egg production.

Chick crumbs and growers' pellets contain coccidiostats that normally protect against the disease, although a vaccine is also available. Alternatively, it is possible to introduce coccidials to young birds at an early age so that they can build up a lifetime's immunity. If a problem occurs in untreated adult birds, your vet should be able to provide you with a prescribed drug.

Coronavirus

Several types of coronavirus cause disease in poultry, including infectious bronchitis, which results in respiratory problems and kidney damage. Generally speaking, the virus causing infectious bronchitis is airborne, spreads rapidly throughout a flock and can persist in an individual bird for several months. The first signs of infection are a drop in egg production and the appearance of rough, wrinkled shells on any eggs that are laid. It is possible to vaccinate against the various coronavirus diseases, and good biosecurity will help prevent their spread to neighbouring flocks.

Newcastle Disease (NCD, or Fowl Pest)

Along with avian influenza, Newcastle disease is classed as 'notifiable' owing to its extremely contagious nature and potentially high mortality rates, although outbreaks can vary from very mild to highly acute. The symptoms include respiratory and nervous disorders, diarrhoea, lethargy and depression. In young birds, Newcastle disease can cause a severe reduction in growth and a number of secondary diseases.

It is spread by direct contact with secretions and excretions, especially faeces, and also via contaminated feed and water vessels. Because the disease is so contagious, in some countries it is compulsory that birds are vaccinated before being allowed to enter any shows. As with Marek's disease, it might pay to form a combined vaccination group with other poultry keepers to reduce costs.

Marek's Disease

There are several forms of this viral disease, a kind of herpes. The classic form appears as lameness in one leg, with the wing dragging on the ground. Some chickens will limp for a time then appear to recover completely, but this is, unfortunately, rare. Some breeds – such as Sebrights and Barnevelders – appear to be more susceptible, whereas Marans and Sumatras are hardly ever affected.

Stress is often the trigger, and pullets are particularly vulnerable at point-of-lay. A high incidence of parasitic worms and introducing the bird to a new environment can result in onset of the disease, which may hitherto have been dormant.

It is possible to vaccinate against Marek's disease, but obtaining small quantities of vaccine may be a problem for the average poultry keeper. An obvious answer is to buy birds that have already been vaccinated, or that have been reared under a broody hen, as these seem to succumb to the disease only very rarely. Alternatively, consider clubbing together with other poultry keepers in a collective vaccination programme.

CHICKEN SELECTOR

BREED	BREED TYPE	BANTAM	PURPOSE	EGG LAYING	EASE OF KEEPING	SOCIABILITY
AMERICAN DOMINIQUE	Soft-feather. Heavy. Rare.	Yes in US	Dual	✔✔	✔✔✔✔	✔✔✔✔
AMROCK	Soft-feather. Heavy. Rare.	Yes	Dual	✔✔✔✔	✔✔✔✔	✔✔✔✔
ANCONA	Soft-feather. Light.	Yes	Layer	✔✔✔✔	✔✔✔✔✔	✔✔✔
ANDALUSIAN	Soft-feather. Light. Rare.	Yes	Layer	✔✔✔	✔✔✔	✔✔✔
APPENZELLER	Soft-feather. Light. Rare.	Yes in US	Layer	✔✔✔	✔✔✔	✔✔✔
ARAUCANA	Soft-feather. Light.	Yes	Layer	✔✔✔	✔✔✔	✔✔✔
ASEEL	Hard-feather. Heavy. Rare.	Yes	Exhibition	✔	✔✔	✔
AUSTRALORP	Soft-feather. Heavy.	Yes	Dual	✔✔✔✔✔	✔✔✔✔✔	✔✔✔✔✔
BARNEVELDER	Soft-feather. Heavy.	Yes	Dual	✔✔✔✔	✔✔✔✔✔	✔✔✔✔✔
BARBU D'ANVERS/D'UCCLE/WATERMAEL	True bantam.	Bantam only	Ornamental/exhibition	✔	✔✔✔	✔✔✔✔
BOOTED	Soft-feather. True bantam. Rare.	Bantam only	Ornamental/exhibition	✔	✔✔✔	✔✔✔✔
BRABANTER	Soft-feather. Light. Rare.	Yes	Ornamental/exhibition	✔	✔✔✔	✔✔✔✔
BRAHMA	Soft-feather. Heavy.	Yes	Dual	✔✔	✔✔✔✔	✔✔✔✔✔
CAMPINES	Soft-feather. Light. Rare.	Yes	Layer	✔✔✔	✔✔✔✔	✔✔✔✔✔
CATALANA	Soft-feather. Light. Rare.	Yes	Dual	✔✔✔	✔✔✔✔	✔✔✔✔
COCHIN	Soft-feather. Heavy.	Yes	Exhibition	✔✔	✔✔✔✔✔	✔✔✔✔
CRÈVE-COEUR	Soft-feather. Heavy. Rare.	Yes	Dual	✔✔✔	✔✔✔	✔✔✔✔
CROAD LANGSHAN	Soft-feather. Heavy.	Yes	Layer	✔✔✔	✔✔✔✔	✔✔✔✔✔
DERBYSHIRE REDCAP	Soft-feather. Light.	Yes	Dual	✔✔✔	✔✔✔✔	✔✔✔✔
DORKING	Soft-feather. Heavy.	Yes	Dual	✔✔	✔✔✔✔✔	✔✔✔✔✔
DUTCH	True bantam.	Bantam only	Ornamental/exhibition	✔✔	✔✔✔✔	✔✔✔✔
FAVEROLLES	Soft-feather. Heavy.	Yes	Dual	✔✔	✔✔✔✔	✔✔✔✔
FRIESIAN	Soft-feather. Light. Rare.	Yes	Layer	✔✔✔	✔✔✔	✔✔✔✔
FRIZZLE	Soft-feather. Heavy.	Yes	Exhibition	✔✔✔	✔✔✔	✔✔✔✔
HAMBURG	Soft-feather. Light.	Yes	Exhibition/layer	✔✔✔	✔✔✔	✔✔✔
HOUDAN	Soft-feather. Heavy. Rare.	Yes	Exhibition/dual	✔✔	✔✔✔	✔✔✔✔
INDIAN GAME	Hard-feather. Heavy.	Yes	Table	✔	✔✔✔	✔✔
IXWORTH	Soft-feather. Heavy. Rare.	Yes	Dual	✔✔✔	✔✔✔✔	✔✔✔✔
JAPANESE	Soft-feather. True bantam.	Bantam only	Ornamental/exhibition	✔	✔✔	✔✔✔
JERSEY GIANT	Soft-feather. Heavy. Rare.	Yes	Dual	✔✔✔	✔✔✔	✔✔✔✔
LA FLÈCHE	Soft-feather. Heavy. Rare.	Yes	Dual	✔✔✔	✔✔✔	✔✔
LAKENVELDER	Soft-feather. Light. Rare.	Yes	Layer	✔✔✔	✔✔✔	✔✔✔
LEGHORN	Soft-feather. Light.	Yes	Layer	✔✔✔✔✔	✔✔✔✔	✔✔✔✔
MALAY	Hard-feather. Heavy. Rare.	Yes	Exhibition	✔	✔✔✔	✔✔✔
MARAN	Soft-feather. Heavy.	Yes	Dual	✔✔✔✔	✔✔✔✔✔	✔✔✔✔
MARSH DAISY	Soft-feather. Light. Rare.	No	Dual	✔✔✔	✔✔✔✔	✔✔✔✔

BREED	BREED TYPE	BANTAM	PURPOSE	EGG LAYING	EASE OF KEEPING	SOCIABILITY
MINORCA	Soft-feather. Light.	Yes	Layer	✔✔✔✔	✔✔✔✔	✔✔✔✔
MODERN GAME	Hard-feather. Light.	Yes	Exhibition	✔	✔✔✔✔	✔✔✔
NAINE DU TOURNAISIS	Soft-feather. True bantam. Rare.	Bantam only	Exhibition/ornamental	✔✔	✔✔✔✔	✔✔✔
NANKIN	Soft-feather. True bantam. Rare.	Bantam only	Ornamental	✔	✔✔✔✔	✔✔✔
NEW HAMPSHIRE RED	Soft-feather. Heavy. Rare.	Yes	Dual	✔✔✔✔	✔✔✔✔✔	✔✔✔✔✔
NORFOLK GREY	Soft-feather. Heavy. Rare.	Yes	Dual	✔✔✔	✔✔✔✔✔	✔✔✔✔✔
NORTH HOLLAND BLUE	Soft-feather. Heavy. Rare.	Yes	Dual	✔✔✔✔	✔✔✔✔✔	✔✔✔✔✔
OLD ENGLISH GAME	Hard-feather. Light.	Yes	Exhibition/ornamental	✔✔	✔✔✔	✔✔
OLD ENGLISH PHEASANT FOWL	Soft-feather. Light. Rare.	No	Dual	✔✔✔	✔✔✔✔	✔✔✔
ORLOFF	Soft-feather. Heavy. Rare.	Yes	Dual/exhibition	✔✔	✔✔✔✔	✔✔✔
ORPINGTON	Soft-feather. Heavy.	Yes	Dual/exhibition	✔✔✔	✔✔✔✔✔	✔✔✔✔✔
OWL BEARD	Soft-feather. Light. Rare.	No	Dual	✔✔✔✔	✔✔✔✔✔	✔✔✔✔
PEKIN	Soft-feather. True bantam.	Bantam only	Ornamental/exhibition	✔✔	✔✔✔✔	✔✔✔✔
PLYMOUTH ROCK	Soft-feather. Heavy.	Yes	Dual	✔✔✔	✔✔✔✔✔	✔✔✔✔
POLAND	Soft-feather. Light.	Yes	Exhibition	✔✔	✔✔✔	✔✔✔✔
RHODE ISLAND RED	Soft-feather. Heavy.	Yes	Dual	✔✔✔✔	✔✔✔✔✔	✔✔✔✔✔
ROSECOMB	Soft-feather. True bantam.	Bantam only	Ornamental/exhibition	✔✔	✔✔✔✔	✔✔✔✔
SCOTS DUMPY	Soft-feather. Light.	Yes	Dual	✔✔✔	✔✔✔✔	✔✔✔✔✔
SCOTS GREY	Soft-feather. Light.	Yes	Dual	✔✔✔	✔✔✔	✔✔✔
SEBRIGHT	Soft-feather. True bantam.	Bantam only	Exhibition	✔	✔✔✔✔	✔✔✔✔✔
SERAMA	Soft-feather. True bantam.	Bantam only	Ornamental	✔	✔✔✔	✔✔✔✔
SHAMO	Hard-feather. Heavy.	Yes	Exhibition	✔✔	✔✔✔✔	✔✔✔
SICILIAN BUTTERCUP	Soft-feather. Light. Rare.	Yes	Laying/exhibition	✔✔✔	✔✔✔	✔✔✔
SILKIE	Soft-feather. Light.	Yes	Ornamental/exhibition	✔	✔✔✔✔	✔✔✔✔✔
SPANISH	Soft-feather. Light. Rare.	Yes	Layer	✔✔✔✔	✔✔✔	✔✔✔
STYRIAN	Soft-feather. Heavy. Rare.	No	Dual	✔✔✔	✔✔✔✔	✔✔✔✔
SULTAN	Soft-feather. Light. Rare.	Yes	Ornamental	✔	✔✔✔✔	✔✔✔✔✔
SUMATRA	Soft-feather. Light. Rare.	Yes	Exhibition	✔✔	✔✔✔✔	✔✔✔
SUSSEX	Soft-feather. Heavy.	Yes	Dual	✔✔✔✔	✔✔✔✔✔	✔✔✔✔✔
TRANSYLVANIAN NAKED NECK	Soft-feather. Heavy. Rare.	Yes	Table/dual	✔✔✔	✔✔✔✔	✔✔✔✔
TUZO	Hard-feather. True bantam. Rare.	Bantam only	Ornamental	✔✔	✔✔✔	✔✔✔
TWENTE	Soft-feather. Light. Rare.	Yes	Layer	✔✔✔	✔✔✔✔	✔✔
VORWERK	Soft-feather. Light. Rare.	Yes	Dual	✔✔✔	✔✔✔✔	✔✔✔✔
WELSUMMER	Soft-feather. Light.	Yes	Layer	✔✔✔	✔✔✔✔✔	✔✔✔✔✔
WYANDOTTE	Soft-feather. Heavy.	Yes	Dual	✔✔✔✔	✔✔✔✔✔	✔✔✔✔✔
YOKOHAMA	Soft-feather. Light. Rare.	Yes	Ornamental/exhibition	✔	✔✔	✔✔✔
YURLOV CROWER	Soft-feather. Light. Rare.	Yes	Dual	✔✔✔	✔✔✔✔	✔✔✔✔

GLOSSARY

Addled A fertile egg, the embryo of which has died during incubation.

Air sac The air space found at the broad end of the egg. It denotes freshness and, during incubation, the development of the embryo.

Alektorophobia The fear of chickens.

Auto-sexing Used of breeds in which the sex of day-old chicks can automatically be determined by the markings on their down. See also 'Sex linkage'.

Axial feather When the wing of a chicken is opened out, you will see along the edge, a single layer of feathers sticking out. Roughly in the middle is a shorter feather, and this is the axial feather, showing where the primary and secondary feathers begin and end.

Beard See 'Muff'.

Blood spot Seen in the laid egg; its appearance could be due to generic, nutritional causes or the hen experiencing sudden stress.

Bloom The moist, protective layer on a newly laid egg. Also, a description of the feathers of a bird in prime and/or exhibition condition.

Booted Having feathers on the feet and legs.

Boule Type of feather formation found on the necks of some continental breeds.

Brassiness Discolouring in light-coloured breeds, caused by sunlight and weathering.

Broiler A young, tender chicken specifically intended for the table.

Brooder An artificial heater used for rearing young chicks.

Broody As an adjective, having the natural instinct to sit on eggs once a clutch has been laid. As a noun, a hen bird showing the same inclination, which will sit either on her own eggs or on those substituted.

Caecum One of two intestinal pouches found at the junction of the small and large intestines.

Candling Using a light to examine the contents of an egg without breaking the shell, often to see if embryos are still alive in the egg.

Cape Feathers running from the back of the head down to the shoulders.

Capon A castrated male chicken.

Carrier Either an outwardly healthy chicken that can, nevertheless, pass on disease to other poultry in the flock; or, more obviously, a box in which chickens are transported.

Clutch A batch of eggs that are incubated together; or, all the eggs laid by a hen on consecutive days before she skips a day and starts a new laying cycle.

Coccidiosis A parasitic protozoal infestation.

Coccidiostat A preventative drug used to keep chickens from contracting coccidiosis.

Cock Mature male after its first breeding year.

Cockerel Male bird before it becomes a cock. Also used as a generic term for male chickens of any age, especially in the UK.

Conformation The shape of a chicken; the standards for which of all pure breeds are laid out in books normally published by the poultry clubs of each country.

Crest Tuft of feathers on the heads of some breeds. Sometimes known as a 'tassel'.

Debeaking Trimming back a bird's upper beak to prevent it from feather-pecking.

Droppings board A removeable board fitted under the perches to collect faeces.

Dubbing The removal of the male bird's comb and wattles, first carried out during the days of cock fighting to prevent injury. Until recently, show strains of some breeds – such as the Old English Game – were still dubbed, but the practice is now either illegal or frowned upon in most countries.

Duck-footed A serious show fault where the rear toes are out of line.

Ear lobes The patches of skin below a chicken's real ears. In most cases, the colour of ear lobes indicates eggshell colour: white-lobed birds lay white or cream eggs, while red-lobed birds lay brown or tinted eggs.

Ear tufts Clumps of feathers growing from small tabs of skin usually found at or near the region of the ear openings.

Egg tooth Found on the upper part of a chick that has just hatched and used to chip through the eggshell. It is lost after a few hours.

Embryo The developing chick inside an egg.

Flight feathers The large primary feathers on the last half of the wing.

Fold unit Portable combined house and run.

Force moult Artificially induced to moult at an unnatural point of a bird's cycle. Sometimes carried out to ensure a bird is in prime condition for a show on a given date.

Free-range Access to the outdoors, sometimes restricted, sometimes unrestricted.

Frizzled Where each feather curls up so that its tip points towards the bird's head.

Fryer See 'Broiler'.

Gizzard A grinding stomach with a muscular lining.

H5N1: A particular strain of avian influenza (bird flu).

Hackles See 'Cape'.

Heavy breed A breed whose ancestry possibly derives from Chinese fowl.

Hen Female after her first laying season.

Immunity Resistance to a particular disease.

Impaction Blockage of a particular part of the body, normally used in connection with problems appertaining to the crop.

Juvenile The first set of proper feathering to moult out before the adult feathers appear. In showing, a juvenile category refers to the age of the exhibitor, not the age of the chicken.

Keel The bony ridge of the breastbone.

Light breed A breed that is usually Mediterranean in origin or has jungle fowl ancestry.

Litter Covering for the floor and/or nest boxes. Normally wood shavings or straw, but can be dried leaves, shredded paper, flax or any clean absorbent material.

Mille fleur A colour pattern, not a breed.

Mite Tiny parasitic insect that lives upon the skin and feeds from the body.

Meat spot Small deposit of blood in an egg, caused by the rupture of a blood vessel during formation.

Moult The period when a chicken sheds its old feathers and grows new ones.

Muff Puff of feathers on either side of the face. Usually seen in conjunction with a beard.

Neck moult Moulting of the neck feathers only; it often occurs when a pullet starts laying before she is fully mature.

Oil sac A gland at the base of the tail that contains oils used in preening to keep the feathers in good condition.

Oviduct The tube down which an egg travels when it is ready to be laid.

Parasite Any insect that feeds on the body of a host, taking nutrition and shelter from the host or its immediate surroundings such as the end of perches and dark corners of nest boxes.

Pigmentation The colour of a chicken's beak, legs and vent.

Parson's nose The protruding lump of flesh normally covered by the tail feathers and seen only on a plucked carcass.

Point-of-lay (POL) The time at which a hen bird is expected to commence laying, usually anywhere between the ages of 18 and 22 weeks, although some breeds will start laying later than this.

Primary feathers The long, stiff feathers at the outer tip of the wing.

Pullet Young hen from hatching to the end of the first laying season.

Rust A show fault, denoting an undesirable reddish area on the wings of partridge, duckwing or pile-coloured females.

Saddle Part of the chicken's back just before the tail.

Scales The covering on the legs and feet.

Scratch Grain fed to chickens, normally as an afternoon feed.

Secondary feathers The feathers on the outer side of the wing situated between the primaries and the point at which the wing joins the body.

Setting The act of putting fertile eggs under a broody hen or in an incubator to hatch them.

Sex linkage When the sex of chicks can be distinguished at the time of hatching by their appearance. If, for example, a gold male is mated to silver females, the pullet chicks will follow the colour of the father while the cockerel chicks will be almost wholly silver.

Shank The part of the leg between the claw and the first joint.

Sickles Long, curved feathers on the outer side of a cock bird's tail.

Spur Pointed, horny projection at the base and rear of a cock bird's legs. Contrary to popular opinion, the length of the spurs is not a reliable indicator of a bird's age.

Standard Either conforming to the standard i.e. perfect specimen of the breed as laid down by the various poultry clubs; or 'standard-sized' – large fowl as opposed to bantams.

Starve-out Failure of newly hatched chicks to eat after absorbing the egg sac.

Sternum See 'Keel'.

Strain A group or flock of chickens carefully bred over several generations by an individual fancier.

Topknot See 'Crest'.

Trachea The windpipe, which forms part of the respiratory system.

Type The typical shape and colour of a chicken indicating the actual breed.

Undercolour The colour of the fluff and lower parts of the feathers.

Variety A particular type of a pure breed – bantam Partridge and large Whites are, for example, both varieties of the Wyandotte breed.

Vent The orifice at the rear of a bird, also called the cloaca, through which droppings and eggs are passed.

Wattles The fleshy parts immediately below the head.

Wing clipping Clipping the primary and secondary feathers of one wing to unbalance a bird and so prevent it from flying.

INDEX

PICTURE CREDITS